New product development in textiles

The Textile Institute and Woodhead Publishing

The Textile Institute is a unique organisation in textiles, clothing and footwear. Incorporated in England by a Royal Charter granted in 1925, the Institute has individual and corporate members in over 90 countries. The aim of the Institute is to facilitate learning, recognise achievement, reward excellence and disseminate information within the global textiles, clothing and footwear industries.

Historically, The Textile Institute has published books of interest to its members and the textile industry. To maintain this policy, the Institute has entered into partnership with Woodhead Publishing Limited to ensure that Institute members and the textile industry continue to have access to high calibre titles on textile science and technology.

Most Woodhead titles on textiles are now published in collaboration with The Textile Institute. Through this arrangement, the Institute provides an Editorial Board which advises Woodhead on appropriate titles for future publication and suggests possible editors and authors for these books. Each book published under this arrangement carries the Institute's logo.

Woodhead books published in collaboration with The Textile Institute are offered to Textile Institute members at a substantial discount. These books, together with those published by The Textile Institute that are still in print, are offered on the Woodhead web site at: www.woodheadpublishing.com. Textile Institute books still in print are also available directly from the Institute's web site at: www.textileinstitutebooks.com

A list of Woodhead books on textiles science and technology, most of which have been published in collaboration with the Textile Institute, can be found towards the end of the contents pages.

Woodhead Publishing Series in Textiles: Number 105

New product development in textiles

Innovation and production

Edited by
L. Horne

The Textile Institute

WOODHEAD
PUBLISHING

Oxford   Cambridge   Philadelphia   New Delhi

Published by Woodhead Publishing Limited in association with The Textile Institute
Woodhead Publishing Limited, 80 High Street, Sawston, Cambridge CB22 3HJ, UK
www.woodheadpublishing.com

Woodhead Publishing, 1518 Walnut Street, Suite 1100, Philadelphia,
PA 19102-3406, USA

Woodhead Publishing India Private Limited, G-2, Vardaan House,
7/28 Ansari Road, Daryaganj, New Delhi – 110002, India
www.woodheadpublishingindia.com

First published 2012, Woodhead Publishing Limited
© Woodhead Publishing Limited, 2012
The authors have asserted their moral rights.

British Library Cataloguing in Publication Data
A catalogue record for this book is available from the British Library.

Library of Congress Control Number: 2011939657

ISBN 978-1-84569-538-5 (print)
ISBN 978-0-85709-519-0 (online)
ISSN 2042-0803 Woodhead Publishing Series in Textiles (print)
ISSN 2042-0811 Woodhead Publishing Series in Textiles (online)

The publisher's policy is to use permanent paper from mills that operate a sustainable forestry policy, and which has been manufactured from pulp which is processed using acid-free and elemental chlorine-free practices. Furthermore, the publisher ensures that the text paper and cover board used have met acceptable environmental accreditation standards.

Typeset by RefineCatch Limited, Bungay, Suffolk
Printed by TJI Digital, Padstow, Cornwall, UK

Contents

Contributor contact details

(* = main contact)

Editor

Lena Horne, Ph.D.
Associate Professor
Department of Textile Sciences
University of Manitoba
35 Chancellors Circle
Winnipeg, Manitoba
Canada
R3T 2N2
E-mail: horne@cc.umanitoba.ca

Chapters 1 and 2

Professor Steven Frumkin,*
   Professor Samuel Bradley and
   Professor Marcia Weiss
Philadelphia University
School House Lane and Henry
   Avenue
Philadelphia
Pennsylvania 19144
USA
E-mail: frumkins@philau.edu;
   bradleys@philau.edu; weissm@
   philau.edu

Chapter 3

Michael Starbuck
Ctext FTI Textile Consultant
Earl Shilton
Leicester
LE9 7HY
UK
E-mail: mail@mikestarbuck.com

Chapter 4

Sharon Evans-Mikellis
Faculty of Design and Creative
   Technology
A.U.T. University
St Paul Street
Auckland 1020
New Zealand
E-mail: sharon.evans-mikellis@aut.ac.nz

Chapter 5

Professor Steven Frumkin*
   and Professor Marcia
   Weiss
Philadelphia University
School House Lane and Henry
   Avenue
Philadelphia
Pennsylvania 19144
USA
E-mail: frumkins@philau.edu; weissm@
   philau.edu

Chapter 6

Ms Jenna M. Eason
College of Textiles
North Carolina State University
2401 Research Drive
Raleigh, NC 27695
USA
E-mail: AutoTexDesign@gmail.com

Chapter 7

Dr Fianti Noor-Evans*
KPMG – R&D Incentives
147 Collins Street
Melbourne
Victoria, 3000
Australia
E-mail: n.fianti@qmul.ac.uk

Dr Stuart Peters
School of Engineering and Materials
    Science
Queen Mary University of London
UK

Dr Natalie Stingelin
Department of Materials
Imperial College
London
UK

Chapter 8

Professor Dr Alexander Büsgen
Niederrhein University of Applied
    Sciences
Mönchengladbach
Germany
E-mail: Alexander.Buesgen@
    hs-niederrhein.de;
    Alexander.Buesgen@gmx.de

Chapter 9

Dr Patricia Wilson* and Justyna
    Teverovsky
Fabric Works, LLC
Arlington
MA
USA
E-mail: tricia@alum.mit.edu;
    justyna@alum.mit.edu

Chapter 10

Frank T. Piller* and Evalotte
    Lindgens
TIM Research Group
RWTH Aachen University
Kackertstrasse 15
52072 Aachen
Germany
E-mail: piller@tim.rwth-aachen.de;
    lindgens@tim.rwth-aachen.de

Chapter 11

Lena Horne, Ph.D.*
Associate Professor
Department of Textile Sciences
University of Manitoba
35 Chancellors Circle
Winnipeg, Manitoba
Canada
R3T 2N2
E-mail: horne@cc.umanitoba.ca

Bernard Rose
TransTex Technologies Inc.
34051–18 baul Casavant West
St. Hyacinthe, Quebec,
Canada
J2S 0B8

Woodhead Publishing Series in Textiles

Introduction

Toward the end of the 1990s, professionals from many sectors attempted to speculate on many aspects of life in the twenty-first century. In an article entitled 'The importance of clothing science and prospects for the future', published in the *International Journal of Clothing Science and Technology* (2002, 14(3–4): 243–244), Masako Niwa wrote:

> At the turn of the millennium, we must question the basic expectations of technology. As new technologies can have a great impact on industry and economy, much is expected of technology. Society expects economic results from technology. Ought not the field of textile technology to change its direction to concentrate on meeting, through new inventions and discoveries, the most important and essential needs, such as widening our views of the world, creating new cultures, protecting our health, keeping us safe, and raising the quality of our daily lives and welfare?

Now, ten years into the twenty-first century, Masako Niwa's aspirations for textile technology are becoming a reality.

New product developments in textiles have indeed widened our views of the world. The ever-growing spectrum of textile products for medical and health end users has created awareness of the implications of aging populations in various regions of the world. Wars and natural disasters have heightened our sensitivity to safety and protection of people and structures. The need for protective systems for the military has stimulated fruitful research and development into materials that are light in weight but durable, materials that form an effective barrier to block chemical or biological agents, or finishes that render materials less detectable. Natural disasters remind us of the need for safe structures. Textiles are being used to reinforce structures or to form barriers to protect properties and structures from the destructive force of rising waterways, wind and erosion. In man-made disasters such as oil spills, textiles play a role in environmental remediation.

In the twenty-first century, some countries will face the challenge of renewing their aging infrastructure; still more countries will be developing new infrastructure as they experience economic growth. These developments will, inevitably, stimulate a surge of demand for innovative technical textiles. The awareness of the impact of industrial activities on the environment has propelled governments to develop and implement policies for their industrial sectors. The environmental impact of producing textiles is already well known. The 'green' movement and the consumers who support it are encouraging textile scientists and engineers to develop appropriate processes and technologies to reduce the environmental footprint of textile production.

While the ability to develop and design innovative textiles and textile products is essential to the sustainability of textile industries in industrialized countries, the migration of textile production from high-income countries to countries that enjoy competitive advantage in terms of production cost has offered many valuable lessons. Textile and textile product production have continued to be effective engines of growth for developing economies. The same phenomenon has brought attention to both the plight and the latent capabilities of some of the least developed countries in the world. It has also rendered developed countries vulnerable when the manufacturing sector loses its strength as a major pillar of their economic growth. The evolution of the global textile landscape has given us an opportunity to become more aware of places, people and the environment that surrounds them.

The wide range of new developments represented in this book signals a paradigm shift. Textiles are no longer mere inputs into a finished product; they have become sources of solutions to issues that affect society. As textiles are being used with increasing frequency to create new products that serve very specific functions, this phenomenon calls for new business models, interdisciplinary collaboration, and new measures of textiles and product performance. As a corollary, there is a pressing need for critical examination of the manner in which higher educational institutions design and deliver textiles programs.

Finally, not only have the contributors to this book shared their expertise, they have also offered deeply meaningful reminders of the immeasurable value of textiles to the human condition.

Lena Horne

Part I
General overview of innovation and textile product development

1
Innovation and new product development in textiles

S. FRUMKIN, S. BRADLEY and M. WEISS,
Philadelphia University, USA

Abstract: This chapter reviews the nature of innovation, particularly the concept of 'disruptive innovation' and the forces driving market change. It then discusses ways companies can innovate, before specifically considering the nature of innovation in the textile industry. The chapter concludes with a series of case studies of how both larger and smaller textile companies have successfully developed innovative products.

Key words: innovation, competition, corporate strategy, competitive advantage, globalization, textile innovation.

1.1 Introduction: incremental change versus disruptive innovation

The concept of innovation is not new to the marketplace. A review of the history of civilization shows many changes in technology, design, markets and marketing, distribution and business structure. Innovation to meet these changes is expensive; in healthy economic conditions it prospers whilst in difficult times it moves to the back burner of tactical corporate strategy. In the field of textiles, innovation has resulted in a wide range of natural and synthetic fabrics that are lighter, smarter, multi-functional and with a wider range of engineered properties.

Over-arching terms in the field of innovation have been used by many business writers, academicians and industrialists to define the process of innovation. A number of years ago Harvard Professor Clayton M. Christensen wrote *The Innovator's Dilemma* (1997), a book followed by a number of others that redefined innovation for other educators, students and the business community at large. Christensen's ideas on innovation help to explain why successful, competently managed companies can trip up even when they are in tune with their customer base and invest in leading technologies.

Christensen called changes that seep into the marketplace as continual product and process renewal 'incremental change', such as the introduction of individual new fibers, yarns and fabrics in the textile industry. Such an example would be the use of stretch yarn, from the early introduction of Spandex (an anagram of the word expands), to Lycra or elastane. Lycra, invented by DuPont chemist Joseph Shivers in 1959, is stronger and more durable than rubber and is known for its exceptional stretch and recovery properties (elasticity). These new fibers allowed

3

companies to extend existing product lines and applications. Many companies have proved adept at anticipating and making these incremental changes.

Christensen contrasts incremental change with 'disruptive innovation'. This happens when what Christensen calls disruptive technology enters the marketplace, often developed by a new player unbeknown to the leading companies. Disruptive technologies create a new value proposition in the consumer's mind that overturns the perceived value of existing products. An example in textiles is the introduction of man-made fibers in the last century. Their introduction transformed a market-place that had been dominated by natural fibers. It led to a completely new generation of fibers and applications as well as an entirely new set of textile companies in the market.

The concept of disruptive innovation has changed the basic concepts of strategy. Strategy is traditionally rooted in supply and demand conditions, and in concepts such as market share and competitiveness against existing rivals in the marketplace. This concept of strategy no longer applies when disruptive innovation makes both an existing company and its competitors irrelevant. Traditional concepts of strategy need to be replaced by a concept such as 'blue ocean' strategy proposed by Kim and Mauborgne (2005). This strategy suggests that companies can create a space in the marketplace that did not previously exist, a blue ocean, in contrast to more traditional companies operating in an established market, the red ocean. The red ocean is everything that currently is in existence.

These concepts of disruptive innovation and the need for a blue ocean strategy are endorsed by key figures in business today, such as Lou Mulkern, editorial director at DBA Public Relations in New York, one of the country's premier PR agencies specializing in consumer electronics. He has been involved in the high-tech PR business for more than 25 years as both a journalist and PR executive. He has worked with global companies such as Toshiba, TDK, Amazon, Newegg (the second-largest online-only retailer in the US), and others to help refine and communicate their key messages.

Mulkern states that in the PR and media business, people are fond of saying things such as 'perception is reality'. However, in press releases for his clients, touting their new products or services, Mulkern maintains that he always tries to resist the urge to use words such as 'innovation' and 'innovative', so as not to dilute their impact; that is, unless the products or services truly live up to the high standard of innovation. 'Innovation is actually a very specific quality,' he notes, 'and it should be reserved for things that really offer people something new and exciting in their lives' (Mulkern, 2009). Mulkern links innovation with leadership. A company that truly innovates, either by creating 'first-ever' type products or providing ahead-of-the-curve services, is, by definition, also a company that exhibits leadership. In this way, leadership and innovation are really inseparable. From a PR perspective, both are very highly sought after qualities for businesses.

True innovation of this kind is hard to define. In this respect Mulkern refers to US Supreme Court Justice Potter Stewart, who famously attempted to explain the meaning of pornography in 1963 with the words, 'I shall not today attempt further to define it ... but I know it when I see it' (Concurring, *Jacobellis v Ohio*, 378 US 184 (1964)). Innovation of this kind is about changing the rules of the game. An example is a manufacturer such as Toshiba and a product such as the DVD. With its combination of dramatically increased functionality and convenience, this new technology automatically rendered existing technologies (such as VHS) obsolete and created new possibilities in the market.

One of the companies that Mulkern has worked with that he states really takes innovation to heart, is Toshiba, whose corporate slogan is, in fact, 'Leading innovation'. He notes that Toshiba's commitment to innovation goes back over a hundred years. One of the company's founders, Hisashige Tanaka, is revered in Japan as a quintessential inventor and innovator. He became famous for creating intricate mechanical dolls, as well as a perpetual or '10 000-year' clock that is still on display at the Science Museum in Tokyo. He also built Japan's first working model of a steam locomotive. The company he founded in 1875, Tanaka Engineering Works, manufactured electric bulbs, cables, prototype telephones, industrial machinery and other products, later becoming today's Toshiba. The company's many innovations include the release of Japan's first rice cookers in 1955 and the DVD in the mid-1990s, as well as breakthroughs in IT and communications, laptops and mobile computing.

Mulkern believes that manufacturers such as Toshiba, who are serious about innovation, need, like Tanaka Hisashige, to be endlessly inventive with a fascination for what technology can do. More than that, however, they need to be focused on how developments in technology can make life easier, more fun or more efficient. Innovation without a clear understanding of how the design of a product and its capabilities will benefit consumers – like that original rice cooker – will simply not be adopted by consumers.

This interest in, and willingness to embrace, the new means that innovative companies need to adapt and change continually. Some of the most durable and successful companies today are nothing like they were when they were founded. Take, for example, IBM (International Business Machine), 3M (Minnesota and Mining) or even retailers such as the Dayton-Hudson Corporation, now operating as Target.

The International Business Machine Company (IBM®) was founded in 1896 as the Tabulating Machine Company, a company that produced one of the first generation data processing machines. Today IBM is one of the world's most valuable brands, second only to Coca-Cola, and the world's fourth largest technology company. With 400 000 employees worldwide, IBM is the second largest (in market capitalization) and the second most profitable information technology and services employer in the world according to the Forbes 2000 list, with sales of greater than 100 billion US dollars.

The Minnesota Mining and Manufacturing Company was founded in Two Harbors, Minnesota in 1902, later becoming 3M™. The company's roots were in

1.1 3M™ products (courtesy of the 3M™ Corporation).

mining stone from quarries for use in grinding wheels. Today they are known for a wide range of innovative products that began with their first exclusive product: Three-M-ite cloth. Today the company has over 76 000 employees that produce over 55 000 products. 3M has operations in more than 60 countries and its products are available in more than 200 countries (Fig. 1.1).

The Target Corporation (Target) was founded in Minneapolis, Minnesota in 1902 as the Dayton Dry Goods Company. It became one of the pioneers of discount merchandising, opening its first Target discount store in 1962. In 1966 it branched out into discount book selling. Through growing expertise in sourcing and buying in bulk, strong financial control and acquisitions, the Target subsidiary grew to dwarf and absorb its parent company Dayton Hudson, becoming the Target Corporation. Target today is the second largest discount retailer in the US, behind Walmart, ranking 28th on the Fortune list of 500 leading US companies. Among other achievements, it has become the largest hardcover book seller in the US. The company now provides design services (Target Commercial Interiors), has a major online presence and manages a global private label business. Target retail stores are typically spacious, feature in-store dining facilities and include grocery departments, e-trade locations, pharmacies, pre-packed deli items and a wide range of other product lines in well-designed, shopper-friendly environments. Target has traditionally been more successful in the field of affordable fashion than its rival Walmart, which has had to continually replace its design and merchandising team.

1.2   Forces for innovation

One of the most famous analyses of market dynamics was developed by Michael E. Porter of the Harvard Business School in 1979 (Porter, 1979). Porter drew upon industrial organization economics to derive five forces that influence the marketplace:

- buyers
- suppliers
- new entrants
- substitute products
- rivalries.

These five forces determine the competitive intensity and, therefore, the attractiveness of a particular market. This approach helps explain some of the drivers of innovation, whether the changing needs of consumers (buyers), the power of other partners in the supply chain, the emergence of new players, the development of technologies that might render a company's existing products obsolete, or the ability of other companies to replicate and surpass their rivals' products or services.

Most companies use this analysis for incremental price, cost, distribution, quality or product enhancements. However, as Porter's analysis suggests, a market can be transformed by new players and/or substitute products. This reinforces the importance of true innovation to break away to gain a leadership role in a given market. This view of the industry allows for the elimination of market borders and definitions, and can truly lead to added value and growth.

The Gap, for example, was a leading specialty retailer for many years. Other companies then began to copy its products and services. The company failed to continue to innovate, gradually losing its competitive edge and market position. The result was a spiral of losses and the eventual collapse of the company. The Gap could have created a new space for itself rather than fall victim to increased competition (Schmall, 2007). In contrast, other companies have succeeded by being truly innovative, whether Walmart with its superior supply chain expertise, which allows it to drive down prices, (Basker, 2007; Hicks, 2007), or Apple with its innovative products that manage to make rival products obsolete. This disruptive innovation paradigm is what is sought by companies to keep them ahead of the game.

A more recent force for innovation is globalization. In a global market consumers have the ability to select products from around the world, increasing competition. This open competitiveness across borders allows the best products to gain market share. In this instance it is innovation on a global scale that drives the market. Fair, open, innovative and global competitiveness drives innovation. The term globalization also refers to the movement of manufacturing, information and technology across borders (Behnisch, 2006; Tabor and Maniam, 2010). This phenomenon has been epitomized by the multinational company, able to source its raw materials and base its production operations in cheaper, developing countries, whilst maximizing prices and market reach in more affluent, developed markets.

The question of who benefits from globalization is often debated. Innovative products are now often designed in one location, the raw materials sourced in other countries, the parts manufactured elsewhere and assembly undertaken in yet another country or countries. At its best globalization benefits customers, who get good quality products at competitive prices, company shareholders, the high-tech design and other staff employed in the more developed countries, whilst giving employment to workers in developing countries processing the raw materials and manufacturing key components.

The growth of multinationals and the globalization of their impact is wrapped up with the rise of the brand. The astronomical growth in the wealth and cultural influence of multinational corporations over the last fifteen years can arguably be traced back to a single, seemingly innocuous idea, developed by management theorists in the mid-1980s, that successful corporations must primarily produce brands, as opposed to products: 'brand builders are the new primary producers in our so-called knowledge economy' (Klein, 2001). Nike, Levi's, Coca-Cola and other major companies spend huge sums of money in promoting and sustaining their brands. One strategy is to try and establish particular brands as an integral part of the way people understand, or would like to see, themselves.

An example of successful branding is Adidas, one of the leading sports brands in the world, with a broad and unique product portfolio spanning apparel and footwear for professional athletes, to authentic streetwear and premium fashion apparel. The global brand, headquartered in Germany, currently focuses on five global priorities: football, running, training, basketball and the Originals label, positioned as 'the iconic sportswear brand for the street'. Recognized by its distinctive trefoil logo, Adidas Originals has many dedicated retail locations around the world. Originals is the category through which Adidas has reintroduced or reinterpreted many of its most recognizable 'heritage' products, such as the white-and-green Stan Smith tennis shoe. But it is also intended to meet the fashion and style needs of today's pop-consumed, trend-savvy 14–25-year-olds.

In early 2010 Adidas Originals launched a global, cross-media campaign designed to reinforce its image. The brand was seeking to establish an identity that was clearly aligned with its youthful consumer base through original audio-visual content and tone of voice, but that was only half of the battle. Adidas also needed a home base from which it could display and distribute all its exciting new content. What's more, it was looking to build a network of appropriate consumers made up of its target group to become ambassadors for Originals throughout their spheres of influence. Adidas Originals settled on building this hub with its Facebook page, the free public profile that enables companies to share their business and products with consumers on an ongoing basis. One of the company's main goals was to increase the number of people who connected to its page.

According to Chris Barbour, Adidas' global head of digital marketing for the sport style division, Facebook was the ideal place to market the brand to a teenage audience (Bloomberg News, 2009). The brand campaign that Adidas Originals launched in early 2010 was intended to reinforce its image and identity through a series of content pieces, launched episodically each month on YouTube, blogs and other online media outlets, as well as Facebook. The campaign ran in 11 countries: Brazil, France, Germany, Hong Kong, Italy, Korea, the Netherlands, Spain, Taiwan, the UK and the US. Each piece of content was created to highlight and support a particular collection, including the Originals Star Wars collection (shoes inspired by the classic science fiction film) in January, the Augmented Reality collection (shoes with code in their tongues that unlocks access to interactive experiences) in February, and its Women's collection in March.

Over the course of the Originals brand campaign between January and June of 2010:

- 228 928 users connected to the Adidas Originals Facebook page by 'liking it' – a six per cent increase.
- The total number of impressions delivered over the course of the campaign was more than 343 million, with more than 15 million of them being organic impressions.
- Adidas Originals video ads generated more than 722 000 direct engagements.
- As of early September nearly 3.9 million people had connected to the Adidas Orignals Facebook page, making it one of the top 20 brand pages on Facebook.
- On the day of a reach block in March, traffic to the Adidas website increased by 139%.

The ability that Facebook provides to target consumers with a message and hold on to them as committed brand ambassadors is unlike anything that the world of advertising has ever seen. Adidas has seen a steady increase in Originals store traffic over the course of its presence on Facebook with many customers noting interest in a particular product because they 'saw it on Facebook'.

1.3 Organizing for disruptive innovation

Many companies are organized to look for the next incremental change because such changes are easier to identify and deal with. They are safe and maintain the status quo. Change is anticipated and planned for, often by senior management who direct junior staff to investigate particular areas for improvement. Put simply, the organizational structure is always working towards the next incremental change.

Disruptive firms, on the other hand, know that true innovation is hard to recognize, anticipate or plan for. They develop a more flexible (but still structured) approach to product development. The structure lies in allocating resources for innovation, but allowing a degree of freedom to investigate a range of options. Decision-making is often decentralized, allowing more junior staff the initiative to suggest and pursue new ideas. What the company does have is an active product development process.

Published by Woodhead Publishing Limited, 2012

A look at the WL Ross International Textile Group (ITG) shows how innovative thinking gave birth to an innovative company. The foundation of this company was the merging in 2004 of Cone Mills and Burlington Industries. Cone Denim, established in 1891, was a leading supplier of denim to jeanswear and denim-related apparel brands. The Cone brand has been synonymous with innovation, authenticity, quality and service offering unique products from vintage premium to mainstream denims. Burlington provided high style, high-performance solutions for leading brands of menswear, womenswear, activewear, cotton casuals, tailored, uniform and barrier products. Fabric initiatives included blended and performance synthetics, worsted wool, cottons and cotton-blends.

ITG was formed in 2004 by Wilbur Ross, an investor in troubled companies, to consolidate these leading textile and fabric manufacturers and to take forward the strategic vision of repositioning the US textile industry by leveraging its marketing and textile know-how on a global basis. Following the strategy set by its CEO Joe Gorga, ITG is focused on meeting its customers' needs in a dynamic, global environment and providing new solutions that add value and enable the success of its customers. As an example, ITG created the Automotive Safety Group in early 2007 with the acquisitions of Safety Components International and BST Safety Textiles, leading producers of airbags, airbag fabrics and other technical fabrics. ITG is a corporation that has shown the way for innovation in corporate sustainability through both Ross' and Gorga's strategic vision. Ross, the entrepreneur with the financial and corporate vision, and Gorga, with a solid textile background, have seen how understanding of a company's core values and day-to-day mission can survive in the global marketplace. ITG currently operates five primary business units: Automotive Safety, Cone Denim, Burlington WorldWide (apparel fabrics), Burlington House (interior fabrics) and Carlisle Finishing. The company employs approximately 12 000 people worldwide with operations in the United States, Mexico, China, Germany, Romania, the Czech Republic, Poland, South Africa, Nicaragua and Vietnam (Business Week.com, 2010).

Creating fabrics that have high performance characteristics is what makes ITG so unique, with fabric properties such as anti-microbial, anti-static and clean-room barrier fabrics, breathability, colorfastness, fire resistance, insect repellency, k-match color matching, moisture management, stain resistance, stretch quality in worsted wools, superior denim stretch and recovery, uv protection, water repellency, and wrinkle free/wrinkle resistant properties. In addition the company is the trademarked products owner of Armor™, Bodyshield™, Coldblack™, Cocona®, Durepel Plus®, EasyWool™, bwwFusion™, Glide™, MCS® Blocker, Minerale™, No Fly Zone™, PBI Matrix™, SGene™, Sigma 4 Star, Versatech® and WeatherMax™.

How do companies such as ITG sustain this degree of innovation? Workplace environments within the corporate structure can create spaces for free exploratory thinking. Companies can create new institutions out of their existing organizations that can be models for ways to solve problems. This can also be done by bringing in new people from different companies or organizations, for example, with fresh ideas or ways of thinking. Some companies even acquire other companies so that they can tap into their expertise. Smaller and medium-sized (SME) companies are often set up to develop a new idea and are a natural reservoir of innovation. As an example, in addition to its own research and development resources, ITG draws upon an extensive list of strategic technology partners to deliver the innovations and performance features necessary to lead and compete in today's dynamic marketplaces. Technology partners include Clariant, Cocona, Dow Fiber Solutions, DuPont, Dyestar, Ecology Works, Huntsman, Invista, Lanaficio Alfredo Rodina, McMichael Mills, Microban, Nano-Tex, Omnova, Outlast, Performance Fabrics & Fibers, PBI Performance, Premier, Pulcra, Sanitized, Schoeller, Textile Rubber, TSI, Unifi and Vanson.

ITG also participates in a number of leading research groups to expand and leverage resources and technologies throughout our industry and the world. Current partnerships include the Institute of Textile Technology (ITT), American Association of Textile Chemists and Colorists (AATCC), Australian Wool Innovation (AWI), National Textile Center (NTC), Canesis (wool research company headquartered in New Zealand), AgResearch, Hong Kong Polytech, University of Manchester, Gaston College and North Carolina State University. ITG has leveraged its core competencies and its ability to see trends that lead it to develop and partner with concepts and companies, many smaller, that give it both incremental as well as disputative abilities.

Successful innovations in an area such as technical textiles for specific markets are normally outcomes of multidisciplinary efforts – a textile or materials scientist + a microbiologist or chemist + an industrial designer + a pattern maker = the ability to engineer specific products for specific end uses. Hence, in textiles development, one can rarely apply the notion that a single innovative person can manage the process of transforming abstract concepts into tangible textiles that meet the demands of specific markets.

1.4   The textile industry and innovation

In June 1948 there were 1.3 million textile jobs in the US and this industry was one of the largest employers in the country. In 1995 there were half as many textile jobs and today there are less than 200000 jobs according to the US Department of Labor statistics for 2010–2011. This is due to the high costs of production in the US versus those in emerging market countries, particularly in the Far East. To participate in this highly competitive market, US textiles manufacturers have had to find new ways of producing and creating innovative, value-added products that can compete against cheaper commodity textile products from the Far East in particular.

Speaking on the subject of innovation in the textile industry in March 2009, Dr Andrew Dent, Director of the Materials Library and Research at Material

ConneXion, addressed these issues in a talk entitled, 'Innovation in textile functionality' (Dent, 2009). Material ConneXion houses over 3000 different materials and they are continually adding innovative products to their library. The library houses examples of such materials as polymers, glasses, carbon-based materials, materials derived from nature, metals, and ceramics. Materials are displayed for customers to come in and view. Dr Dent pointed out in his talk that it is not only important to create new viable materials but also to innovate in the area of textile design. Materials such as optical fibers, photo and electroluminescents, aero gels, carbon fibers and corn are new and potentially important. A good example is the use of corn to make plastics. Free from the use of petrochemicals, this new innovation could have a significant benefit for the environment.

1.5   Trends in textile innovation: wearable electronics, biomedical, biomimetic and nano-textiles

Technical textiles are a growing field, using technologies such as smart polymers and the convergence of electrochemistry and textiles in order to process electronic polymers into fibers and fabrics. The integration of smart functionality into clothing and other textile products will radically change the culture surrounding these products, fundamentally altering people's relationships with them and the way they use them. Smart functionality will also have an impact on the way products are designed and the materials developed.

1.5.1   Wearable electronics

The integration of electronic technologies in smart textiles will become, according to Baurley (2004), more important than the fashion of the textile clothing products themselves. In this scenario clothes will change their appearance, their tactile quality and even their shapes. Clothing can also serve to reflect, hide or generate mood. By incorporating communication devices or smart cards, clothes will be used as a channel of communication. The Sony Corporation is looking to electronic game developers to connect individuals to make gaming more of a social experience. In this regard Sony and Microsoft are working to replace the joy stick with gestures captured by the wearers'/gamers' movements.

Some currently available or in-development interface textile electronic wearing apparel items include Textronics, a leader in textile sensor technology, which is now offering a do-it-yourself kit for people interested in developing heart rate monitoring textiles. The Developer's Kit offers 12 textile sensors, a variety of knit tubes, wrist cuffs and chest straps. Additionally, they include conductive thread and two transmitters with the kit. These transmitters are compatible with most analog heart rate monitors, such as Polar. This can open the door to many fitness and sporting tech garments.

Published by Woodhead Publishing Limited, 2012

QIO Systems is now offering textile touch pads for fashion fabricators, using Elektex and SOFTswitch technology that can be washed and dry cleaned. Even though it is washable, it is rated at up to 100 000 keypresses. As well as the fabric keypad, QIO Systems offers a wide range of electronic modules with control functions for the iPod, Bluetooth cell phones, wearable radio, push-to-talk solutions and, separately, integrated heating and lighting for garments and soft goods.

Fibretronic has released a new series of controls. The FTXS series is a soft control keypad that is able to be integrated into textiles. The super soft keypad is targeted for mobile devices. This is designed to work directly with Fibretronics CONNECTED-wear modules for iPod, iPhone and MP3. These Fibretronic products are found in many smart textiles currently sold. Fibretronic's Fiddler joystick system, known from the Levi's Redwire Jeans, enables you to upgrade your outfit to work with your iPod. The joystick module comes together with the iPod interface electronics in a blister pack. It allows for easy attachment and removal of the iPod controller to any clothing. The first company to implement this product and create a clothing line is ToBe Technology, a Swedish manufacturer of hip snow sportswear. ToBe's iPod-enabled jackets are designed for easy integration of Fibretronic's iPod control module, giving their customers the flexibility to interchange the control system between different garments in ToBe's product range.

1.5.2 Biomedical textiles

In biomedical research product sampling and experimentation, every implantable product innovation developed involves countless choices. One of the very first is whether or not to consider biomedical fabric structures for use in device components. That is where companies such as Secant Medical, an industry leader experienced in medical textile solutions, use the theory of blue ocean strategy to cross the boundaries of traditional textile use to explore new, untapped markets in developing woven, non-woven, knitted and braided medical textile structures.

1.5.3 Biomimetic and nano-textiles

According to the Biomimicry Institute in the US, biomimicry (from the Greek words *bios*, meaning life, and *mimesis*, meaning to imitate) is a design discipline that seeks sustainable solutions by emulating nature's time-tested patterns and strategies, for example, the Speedo sharkskin swimsuit. The core idea is that Nature, imaginative by necessity, has already solved many of the problems we are grappling with: energy, food production, climate control, non-toxic chemistry, transportation, packaging, and many more.

There have been various innovations in biomimicry, phase change materials, shape memory polymers, structural textiles and three-dimensional textiles.

Published by Woodhead Publishing Limited, 2012

Speedo is a corporation based in Nottingham, England, and is a manufacturer of swimwear and accessories. The original company was founded by Alexander MacRae, as MacRae Hosiery Manufacturers, in Bondi Beach, Sydney Australia. Speedo has been using biomimicry to mimic the surface of sharkskin to reduce friction in water. By creating a swimsuit with less drag in water, they are a cut above the other swimsuit manufacturers. Speedo's Aqualab team is working to continually use technology to be on the cutting edge of sport science innovation, forever creating new dimensions of swimwear, sports equipment and apparel. Working with the leading scientists from varied industries new innovations are developed, trial tested and then partnered with leading athletes, ensuring that Speedo is always ahead of the competition in swimwear technology. The experts that Speedo uses are in the fields of sport science, textile fabric apparel technology and engineering and aerospace.

An example of the use of nanotechnology in textiles is the Alexium Group in London, which has been licensed to use a technology developed by the US Air Force to produce self-cleaning underwear fabric. Already in use by the US military, these fabrics have been used to create t-shirts and underwear that can be worn hygienically for weeks without washing. The new technology attaches nano-particles to clothing fibers using microwaves. Chemicals that can repel water, oil and bacteria are then directly bound to the nano-particles. These two elements combine to create a protective coating on the fibers of the material. This coating both kills bacteria, and forces liquids to bead and run off. This innovation seems to be incremental but may prove to be disruptive if it changes the way consumers use these garments.

1.6   Case studies in innovation in textile manufacture

1.6.1   Milliken and Company

A textile company long known as an innovative leader in its field is Milliken and Company. Founded in 1865 the company became, in the late 20th century, the largest family-owned textile business in the world. That had already been proven when new mills opened in the South in the 1890s. New Southern mills and inexperienced labor soundly beat the New England manufacturers with their older equipment and experienced hands.

In 1944 the US War Production Board issued a Certificate of Necessity for Deering Milliken to manufacture nylon tire cord. Accordingly, a totally new facility was designed and built for one purpose – to manufacture tire cord as efficiently as possible. The Excelsior Tire Cord Plant was a one-story building arranged to facilitate the flow of raw material through all stages of manufacture and delivery to the shipping dock. It was the first textile mill built without windows and with complete air cleaning and cooling systems. The mill set a pattern that would be copied over and over throughout the industry.

Roger Milliken succeeded as president in 1947. He took over the job of running the mills, expanding operations and undertaking research that would take the company to the forefront of textile innovation and implementation. A research team (Deering Milliken Research Trust) had been organized in Clemson in South Carolina in 1945. The first research development of worldwide significance was a patented edge-crimping texturing process for nylon continuous filament yarn. Manufacturing rights were licensed to many American and foreign companies, and Milliken began to be recognized as an innovative and technology driven company. Trade named 'Agilon', the resulting yarn was, for many years, the leading yarn for producing one-sized stretch women's hosiery, due to its non-torque properties.

In 1954 Milliken moved to Spartanburg in South Carolina, a move that stamped his mark on the move to the South that had been underway throughout the century. In 1958 a research park was built in Spartanburg. Deering Milliken Research Trust became Deering Milliken Research Corporation. This new location included a Management Information Center, a chemical pilot plant and a Model Manufacturing Center (called the Prototype Plant). Slowly, Spartanburg became the corporate headquarters for the company. Customers were invited to stay in a guesthouse on the property that had sleeping accommodation for 29 persons and dining facilities for 100. A helicopter landing pad was added to allow easy access to nearby Greenville-Spartanburg Jetport or local flights to nearby mills.

It was in Spartanburg that the second research development of worldwide significance was developed, trade named 'Belfast'. Dmitry M. Gagarine invented and developed a process for 'wet cross linking' of cotton, which was used by both Milliken and its licensees to produce cotton fabrics with wet memory for flat drying, giving them 'drip dry' properties, which remained popular with consumers until the spread of tumble driers in people's homes and the advent of polyester blends from DuPont. Millikin did not only innovate in technology. It also actively sought to innovate in production and in finding new markets to enter and services to sell. An example of the latter was the Kingsley Mill, built in the 1950s to cut and package fabric for retail sale. The plant was named after Francis Kingsley, a Deering Milliken executive who planned and operated the Milliken Breakfast Show, an annual advertising show. The show brought customers to New York each October where a dazzling display, modeled on Broadway productions, featured new fabrics.[1] Innovative in its market appeal, and attended by thousands, tickets to the show were sought after by all major executives of the textile industry.

As an example of innovation in production, in the early 1960s Milliken began to question the conventional thinking of most textile executives regarding inventories. Traditionally, manufacturing and marketing were treated as separate functions, often resulting in excessive inventories. Milliken commissioned a study that indicated an inverse relationship between inventory size and profits. This led Milliken to keep tighter control of inventory by adjusting the rate of production, effectively becoming a pioneer of 'lean' manufacturing.

The name Deering was finally dropped from the company name in 1978. Milliken and Company was born. Innovation continued. Always looking for ways to improve and modernize the company, Milliken launched the 'Pursuit of Excellence' program in 1981. The program emphasized self-directed teams of employees who met regularly to discuss ways to improve the process and the product. As proof of this unique innovation-based company, internationally-known management guru and author Tom Peters dedicated his 1987 bestseller, *Thriving on Chaos*, to Roger Milliken.

A final example of technical innovation is Visa, a fabric finish that resists stains and is used on a wide variety of products, including clothing and tablecloths. The original irradiation process for making Visa has been replaced by a chemical process. The development of Visa strongly reaffirmed the company's position as a leader in developing patented fabric finishes. It also produced huge profits. As a result of such developments, in 2001 the company was awarded the Textile Industries Innovation Award for creating a record of corporate success in innovation that can be traced to a corporate culture that fosters free thinking, idea exchange and continuing education. In corporate circles Milliken and Company is mentioned along with DuPont, 3M, Motorola and Dow Chemical.

1.6.2 DuPont

Founded in 1802, E. I. du Pont de Nemours and Company (DuPont) has consistently sought to implement a corporate strategy of 'using science in its work by creating sustainable solutions essential to a better, safer, healthier life for people everywhere. Operating in approximately 80 countries, DuPont offers a wide range of innovative products and services for markets including agriculture, nutrition, electronics, communications, safety and protection, home and construction, transportation, textiles and apparel' (DuPont, 2010). DuPont produces a wide range of products including synthetic fibers, polymer resins, packaging films, automotive finishes, crop protection products and industrial chemicals. The company serves global markets through a number of subsidiaries in Canada, the United States, Mexico, France, the UK and India.

DuPont's first major product was explosives. Not surprisingly, safety has been a high priority for the company since its inception. According to Rhonda Carlin, DuPont Canada's Business Sustainability Resource, the company would probably not be celebrating its 200th anniversary if it had not started out by emphasizing safety in the manufacture and use of its products, and by instilling safety early on as a core value of the company. DuPont and, in particular, DuPont Canada have some of the best safety records in the world and are committed to evolving their business portfolios while staying true to their safety ethic and values. Over the years the company has transitioned from making explosives to manufacturing industrial chemicals and products, and is now striving for less material-intensive, more knowledge-intensive, products and services.

Published by Woodhead Publishing Limited, 2012

DuPont has been a major textile manufacturer for many years. Beginning with Rayon, a regenerated fiber developed in the late Victorian era, the company moved into developing completely synthetic fiber products from chemicals found in the petroleum industry. Synthetic fibers were developed in the 1930s and 1940s, and came into general use in the 1950s. These new fibers included Polyamide (Nylon), Polyester, Polyacrylonitriles (Acrylics), Polyolefins and Polyurethanes (Spandex and Lycra). Later combinations with natural fibers, such as wool or cotton, introduced the consumer to the concept of easy care fiber blends with a natural feel. Poly/cotton garments and acrylic mixed knits could be produced faster at lower cost, and soon displaced traditional cotton and wool garments. Viscose rayon became popular again in the 1980s, when there was the start of a reaction against synthetic fibers. Viscose crinkle fabrics and fabrics with exceptional drape or sheer effects wooed the consumer back. By 2000 designers had caught on to adding Lycra or Spandex to fibers like viscose and acetate to create garments with greater comfort and better shape retention in wear.

At the Shanghai World Expo in March 2010, DuPont showed how their innovations are able to deliver improved energy solutions and safer, more efficient buildings and interiors, among other applications (DuPont, 2010). Applications include:

- *DuPont Apollo®* Thin-film photovoltaic modules are used for commercial rooftop and large-scale applications. The solar modules can generate more wattage output under diffuse lighting conditions and consume only about 1/200 the silicon of traditional crystalline silicon solar cells, resulting in shorter energy payback times.
- *DuPont™ Tyvek® Weatherization Systems* Part of a system created to seal buildings from the inside and out, Tyvek® is an Energy Star partner that helps enhance the energy efficiency, indoor air quality and overall comfort of a home or commercial building. Tyvek®, which combines properties of paper, film and fabric, is also an ideal choice for reusable bags by environmentally conscious customers.
- *DuPont™ Energain®* Decreasing the amount of energy used in a building by controlling temperature levels, and thus reducing the need for cooling and heating, the results are major cost savings and reduced carbon dioxide (CO_2) emissions.
- *DuPont™ Sorona®* Renewably sourced polymers are made partially with agricultural feedstocks instead of petrochemicals, thus reducing dependence on oil. In addition to fibers and fabrics, Sorona® can be used in films, filaments, engineering resins and other applications. Sorona® contains 37% renewably sourced ingredients by weight.
- *DuPont™ Corian®* Solid surfaces, made from natural minerals and high-performance acrylic, non-porous, and helping to resist stains and the growth of bacteria. Available in more than 100 colors and all fulfilling the US National Sanitation Foundation (NSF) Standard 51, Corian® is safe for food contact.

Published by Woodhead Publishing Limited, 2012

- *DuPont™ SentryGlas®* Ionoplast interlayers help create lighter, safer, more structural glass that can stand up to greater loads and higher threat levels.
- *DuPont™ Teflon® FEP* Anti-flammable cable providing excellent fire resistance characteristics without producing toxic smoke in the event of fire, thus gaining valuable time for evacuation of personnel. In addition, it can be recycled.

A major turning point for DuPont came in 1988 when Greenpeace named the company 'Number One Corporate Polluter' in the US. Greenpeace representatives scaled one of the company's stacks at its New Jersey facility and hung a huge banner that read 'DuPont Number One Polluter' to draw media and public attention to the company. At this point, DuPont asked itself a serious question: 'Even though we comply with environmental laws, is this where we want the company to be?' And the answer was no. The company acknowledged that in spite of its compliance record, it was still a significant polluter and that something could be done. The event marked the beginning of an enormous effort to reduce the company's pollution and reduce its overall environmental footprint. DuPont developed a Global Commitment to Safety, Health and the Environment and laid out a number of actions that would help the company achieve this commitment to reduce its environmental footprint. DuPont Canada also shares the same commitment to global health, safety and the environment. To fulfill it, the company states that it will:

- adhere to the highest standards of performance and business excellence
- aim for the goal of zero injuries, illnesses and environmental incidents
- drive toward zero emissions and zero waste generation
- excel in the efficient use of energy and natural resources, and manage lands to enhance wildlife habitat
- continuously improve processes, practices and products to reduce risk and impact throughout the product life cycle
- promote open and public discussion of environmental issues and build alliances to develop sound public policies and regulations
- educate, train and motivate employees and executives to comply with the environmental commitment, and provide accountability by reporting regularly to the public.

The company's website includes the following global commitment to safety, health and the environment (DuPont, 2010):

- We will conduct our business with respect and care for the environment.
- We will implement those strategies that build successful businesses and achieve the greatest benefit for all our stakeholders without compromising the ability of future generations to meet their needs.
- We will continuously improve our practices in light of advances in technology and new knowledge in safety, health and environmental science.

Published by Woodhead Publishing Limited, 2012

- We will make consistent, measurable progress in implementing this commitment.

DuPont sees value in being a role model for sustainability, and subscribes to and endorses a number of international codes of conduct or principles related to sustainability and corporate social responsibility. For example, the parent company has endorsed the United Nations Global Compact, citing that 'DuPont's core values of innovation and discovery, safety and environmental stewardship, integrity and high ethical standards, and treating people fairly and with respect meet and in many respects exceed the goals embodied in the values set out in the Global Compact'. By addressing a major challenge and, indeed, taking a leadership role in environmental issues, DuPont has kept ahead of competitors and created new market opportunities.

1.6.3 Cotton Incorporated

Jacobson and Smith (2001) have studied textile market innovation driven by the global economy by focusing on the accomplishments of Cotton Incorporated. Cotton production and the number of cotton farmers dwindled rapidly from the early 1970s to the 1980s, when the cotton farming and manufacturing base was threatened with collapse by the rise in the use of synthetic fibers and the growth in cheaper overseas competition. In 1973 US cotton production accounted for only 12% of total global market share. This position can be contrasted with 2000 when the US market share was 19.5%. Jacobsen and Smith attribute the salvation of the US domestic cotton market to the work of Cotton Incorporated.[2] Cotton Incorporated was founded in 1970 by US cotton growers and manufacturers to represent their interests around the world, including the opening of offices in many countries, such as China, in order to promote US cotton. The organization's mission was to market the exceptional characteristics of American cotton fiber and to find new ways of exploiting those characteristics to create differentiated products in the marketplace (Cotton Incorporated, 2010). Three fabric innovations by Cotton Incorporated that have found their way into the market are:

- *Wicking Windows*™, the moisture management system that improves absorbent capacity while reducing the drying time of cotton knits.
- *Stay True Cotton*™ technology, which locks in the original indigo color in denim.
- *Storm Denim*™, which imparts water-resistance while allowing water vapor to pass through the fabric.

New York designer Alexander Wang has applied Storm Denim technology to his premium jeans being introduced at high-end retailers. Canadian MWG Apparel is also applying Storm Denim to a line of premium denim jeans being sold by Mark's Work Wearhouse, one of that country's largest retailers.

Further advances in sustainable finishing from Cotton Incorporated are:

- rapid exhaust bleaching cycles that increase productivity and use less water and energy
- continuous bleaching that combines scouring and desizing and uses less water and energy
- the use of ozone to reduce chemicals in wash-down and to decolorize water so that it is less polluting
- the application of foam to fabrics to transfer dyes, using less water, chemicals and energy.

This innovative approach has helped the US cotton industry lead the market and thus rebuild its position.

1.7   Sources of further information and advice

General

Anon. (1994). '*Textile World*'s 1994 Leader of the Year: Dr Thomas J. Malone', *Textile World*, October, pp. 34–41.

Biomimicry Institute website, http://www.biomimicryinstitute.org/.

Baurley, S. (2004). *Interactive and Experiential Design in Smart Textile Products and Applications*. Cambridge: Woodhead Publishing.

Christensen, C., Horn, M. B. and Johnson, C. W. (2008). Disrupting Class: How disruptive innovation will change the way the world learns, New York: McGraw-Hill.

Cotton Incorporated (2010). http://www.cottoninc.com/.

Dent, A. (2009). 'Innovation in textile functionality', class presentation, University of Rhode Island, Textiles, Fashion Merchandising and Design.

Johnson, M. W., Christensen, C. M. and Kagermann (2008). 'Reinventing your business model', HBR Business Articles, 1 December.

Orr, S. (2006). 'Spartanburg, SC, textile company named one of the country's best employers', *Knight-Ridder/Tribune Business News*, 9 January, 2006.

Whaley, P. (2004). 'Milliken & Company: Covering all bases', *Textile World*, April 2004.

Case studies

McAfee, A., Sjoman, A. and Dessain, V. (2004). Zara: IT for Fast Fashion, 25 June, 23pp. (Prod. #: 604081-PDF-ENG).

Montgomery, D. B., Carducci, E. and Horikawa, A. (1994). *Levi Strauss Japan K.K.* 11 May, 37pp. (Prod. #: M276-PDF-ENG).

Wison, R. E. (2010). *Target Corporation: Maintaining Relevance in the 21st Century Gaming Market*, 2 April, 25pp. (Prod. #: KEL442-PDF-ENG).

1.8   Notes

1 My father was the Executive Director of the Textile Veterans Association (TVA), representing people who had both worked in the textile industry and were also World War II veterans, which raised funds for textile educational institutions and supported veterans' hospitals and families. Milliken and Company was a major contributor to the TVA and both Roger and his brother Minot were good friends of my father, Murray

Frumkin. From this association my first introduction to the textile industry was being invited to the Milliken Show.

2   In the interest of fair disclosure, the author, Steven Frumkin, was awarded a grant by Cotton Incorporated in 2009–2010 to promote awareness and to increase education for those who wish to enter the field of cotton fashion, textiles, apparel and retailing as a career path. This included the Internet radio show, Cotton Radio, which was streamed/broadcast weekly in the fall of 2010 on Voice Americas' Business and Variety channels. The show is available as archived by Voice America, and is downloadable via iTunes.

1.9   References

Basker, E. (2007). 'The causes and consequences of Wal-Mart's growth', *Journal of Economic Literature*, 21(3): 177–198.

Behnisch, A. (2006). 'How to define globalization?', Paper presented at the annual meeting of the International Studies Association, Town & Country Resort and Convention Center, San Diego, California, USA. Available from http://www.allacademic.com/meta/p98917_index.html.

Businessweek.com. International Textile Group, company profile, 2010.

Christensen, C. (1997). *The Innovator's Dilemma: The revolutionary book that will change the way you do business*, Boston, MA: Harvard Business School Press.

DuPont (2010a). 'Learn more about DuPont …', available at http://www2.dupont.com/Our_Company/en_US/.

Hicks, M. (2007). *The Local Economic Impact of Wal-Mart*. Youngstown, NY: Cambria Press, 337pp.

Jacobson, T. and Smith, G. (2001). *Cotton's Renaissance: A study in market innovation*. New York: Cambridge University Press.

Kim, W. C. and Mauborgne, R. (2005). 'Blue Ocean Strategy: How to create uncontested market space and make the competition irrelevant', Boston, MA: Harvard Business School Press.

Klein, N. (2001). *No Logo: Taking aim at the brand bullies*. London: Flamingo.

Porter, M. E. (1979). 'How competitive forces shape strategy', *Harvard Business Review*, March/April.

Schmall, E. (2007), *Who'd Buy The Gap?* New York: Telsey Advisory Group.

Tabor, L. and Maniam, B. (2010). 'Globalization: trends and perspectives of a new age', *The Business Review, Cambridge*, 15(1): 39–45. Retrieved 21 July, 2010 from ABI/INFORM Global (Document ID: 2045078931).

Womack, B. (2009). 'Facebook sees fourfold jump in advertisers since 2009', Bloomberg News.

Published by Woodhead Publishing Limited, 2012

2

Practical aspects of innovation in the textile industry

S. FRUMKIN, S. BRADLEY and M. WEISS,
Philadelphia University, USA

Abstract: The process of taking an innovative apparel or textile concept and turning it into a successful product is challenging and multi-faceted. Innovation is the process of developing a product offering that more effectively satisfies the unfulfilled needs of the target market better than the competition. Furthermore, any emerging technologies must meet industry standards for safety and performance. Social networking and mass customization are effective tools to create a buzz about new products and allow consumers to provide feedback to the apparel firms on the emerging consumer trends and acceptance of their product offering. The intellectual property developed during the innovation process must be protected from competition and imitators.

Key words: trade dress, intellectual property, point of differentiation, mass customization, innovative apparel.

2.1 Introduction and practical aspects of innovation

How does one gauge if a product concept is innovative? From whose perspective is the value of the new product idea judged? New and improved technology becomes innovative when the target market perceives the offering to provide greater value than currently available products. The innovation process is not about developing the most technologically advanced product because technology does not ensure a win in the marketplace. Customers are looking for fulfillment of unmet or underserved wants or needs. A new product must offer a simple, clear and distinctive advantage over what is currently available; this is called the point of differentiation.

The first section of the chapter addresses the importance of innovation from the perspective of fulfilling unserved and underserved needs of the target market, and doing so better than the competition. Innovation is not about providing the latest and most advanced technology, but rather is a process of understanding the wants and needs of consumers and providing products that will be perceived as having more value than existing products. In the marketplace, the most successful products are the ones that can be clearly differentiated from all of the current offerings.

Product differentiation focuses on the benefits of the offering, which is that intangible characteristic that answers the question: what is the advantage to the consumer that is offered by this product, which is unlike any other available on

22

the market? Features are the physical characteristics, such as color, size, shape and material, and any other physical description of the item. Benefits, however, address what advantage will be received by the consumer such as status, luxury and pride of ownership. Marketers generalize about the distinction between feature and benefits by saying: do not sell the steak, sell the sizzle. In a similar way, apparel firms do not sell the textile, they sell beauty. Innovation cells facilitate the creation of products that identify and effectively address the needs of customers.

Innovation cells are teams from a given firm that are formed to generate ideas regarding new products and improvements to existing products. These ideas should come from all levels of the organization as opposed to being driven down from the top. A common misconception exists that innovation is the implementation of the ideas of either senior management or research and development teams. The best innovation cells collect data from all levels of the firm, and pay particular attention to members of the team that have direct contact with consumers and have intimate knowledge about the competitive environment. Before the innovative products are released to the market, the organization must take appropriate steps to ensure that the intellectual property that was created by the innovation cells is properly protected.

Intellectual property provides protection that prevents competitors from stealing print patterns, styles or any other characteristic of a textile or apparel in an attempt to deceive the public regarding the origin of the given product. Legal protection is provided through copyrights, patents and trademark protection. Intangible attributes are protected by trade dress laws.

2.2   Meeting the needs of customers better than the competition

The first step in creating a strong point of differentiation is developing an understanding of the target market, competition and the brand position of the new offering. Customers will view a new and potentially innovative offering in the context of the available products, and determine if the given consumer has undersatisfied or unsatisfied needs that are better fulfilled by the new product. Each consumer will evaluate new offerings in terms of whether or not it outperforms the currently available products and is consistent with the value proposition expected from the given brand (Bowonder et al., 2010). If the answer to those questions is yes, the new offering will have greater perceived value to the customer.

Victoria's Secret is viewed by the firm's target market as a provider of high quality feminine women's wear. In terms of brand equity, the Victoria's Secret brand would have no value for products that are viewed as masculine. The firm would have a difficult time marketing sporting goods equipment for outdoorsmen or ice hockey players. Although it might be innovative to expand into a Victor's

Secret brand of gifts for men (targeted to female shoppers), the brand equity of the current line could be compromised, and the new offerings would be, at best, questionable. The innovation process must create more than additional product features because value creation occurs when the consumer perceives greater benefit.

During the innovation process, the product development team must consciously address the perception of the brand that the firm brings to the marketplace. If the product does not match the value proposition expected from the given brand, then customers will reject the product. Product features are not necessarily an indication of innovation because customers are looking for the 'what's in it for me?' benefit. As an example, a customer who is shopping for workout clothing is typically not concerned about the composition of the fibers in the textiles used to make an Under Armour t-shirt. The buy decision is based on the performance of the product and the perceived value of the Under Armour brand. Under Armour build their premium brand by providing the benefit of high performance, moisture-wicking shirts for athletes. The image was built on performance, and high-quality performance is clearly linked to the brand and the premium price at which the line is offered. Innovation, in the case of Under Armour, was built on providing a unique set of benefits not offered by any other apparel company in the market, not by focusing on the product features of textile materials or other technical specifications. As the firm began to provide product line extensions, such as shorts, pants, hats and jackets, they never strayed far from the brand's core competency.

2.2.1   Features versus benefits

Another key aspect of understanding the practical aspects of innovation is that technology is the not the key to success in the market. Successful new products are better at meeting the needs of the target market more effectively than the competition. Offering all of the latest bells and whistles does not ensure success in the marketplace (How Innovation Really Works, 2010). A great example of competing products where the less technologically advanced offering was the winner in the marketplace was the war between Betamax and VHS for dominance in the video home-recording market. Betamax, by all accounts, was a superior technological system offering superior sound and picture quality. It was also the first to market in 1975, which is typically a distinct advantage because it provides the offering firm the opportunity to position the product as an industry leader. VHS was the later market entrant (unveiled in 1976), offered lower video and audio quality and the tape cassettes were substantially larger (Bulik, 2006).

In the minds of consumers, however, these differences were relatively unimportant when compared to the recording time offered by each system. Betamax initially offered only one hour of recording time per tape, while the VHS system offered two hours of recording time, which provided sufficient time for a movie to be completely taped on one VHS cassette. As the technology advanced, recording time per cassette eventually doubled (and then tripled), VHS maintained its recording-time advantage by allowing two, then three movies to be stored on one cassette. The marketers of the VHS technology addressed the critical point of differentiation of recording time, which was the innovative factor on which consumers placed the most value, and VHS became the standard format for home video-tape recording.

2.2.2 Innovation cells

The previous discussion clearly points out that innovation is not about providing the most advanced technologies, but requires those involved in the innovation process to focus on how value is perceived by current and potential customers. Given that difficult economic times create a very strong competitive environment, today's apparel industry is under extreme pressure to be innovative, but the process is often misunderstood or improperly implemented. One of the most commonly held misconceptions about the innovation process is that new and exciting product ideas come exclusively from either research and development teams or senior managers who are product development gurus (Igartua *et al.*, 2010). Innovation is not product driven, but rather customer driven. The first step in the innovation process is to develop an understanding of the needs of the target market. Those organizations that are the most effective innovators engage a broad range of intra-business disciplines to develop products in innovation cells.

Team members involved in innovation cells have three basic charges: developing products that will excite customers; out-performing the competition by creating relevant strong points of differentiation; and building the product and brand portfolio (Bowonder *et al.*, 2010). The most successful innovation cells are organized to consider ideas and products created at any level of the organization, and should include those members of the firm that have the most direct contact with the customers, that is, the sales, marketing and customer service staff. These are the team members that are the front line soldiers who not only do battle daily with the competition, but can provide direct, unfiltered feedback regarding the positive and negative perception of the ultimate consumer. Innovation cells provide an important framework to address many key elements of the innovation process.

Beyond addressing the needs of the market, an effectively structured innovation cell defines the objectives and goals of the process. Innovation cannot be successfully accomplished in a vacuum, and must incorporate 'a framework for developing a product innovation strategy that includes defining innovation goals and objectives, selecting strategic arenas, developing a strategic map and location resources' (Cooper and Edgett, 2010: 33). When properly configured, innovation teams create a template for the product before the very first design idea is created. Today, the product development process must begin with a clear understanding of the purpose and focus of the innovation process from the customer's perspective.

Self *et al.* (2010) discussed the five key principles that must be addressed as the innovation cell is formed. The first principle is that the team must have the ability to share accurate information. Innovation cannot happen in an environment where the members of the cell must confirm the currently held beliefs, but rather all team members must be able to provide accurate and non-biased feedback relative to the process. Senior management is responsible for providing a clear set of objectives to the innovation cell team members during the formation of the cell.

Active listening is the second key principle to the successful implementation of an innovation cell. All cell members must incorporate not only their data into the process, but also be willing and able to absorb and process the information provided by other team members. As the process moves forward, management must be willing to encourage employees to take risks and support their risk-taking behavior. The most innovative ideas are those that challenge existing paradigms, which requires the innovators to not be risk averse. Risk taking can lead to ideas that may not be successful, so it is important that the innovation cell members use failure as a learning tool. The final critical principle is fostering mutual trust. Everyone involved in the process must be willing to be vulnerable, with the understanding that every member of the innovation cell is working towards a common goal.

2.3   Innovation as a driver of new strategic issues in the apparel industry

One significant challenge facing apparel and textile firms is the protection of the intellectual property. Intellectual property is defined as the intangible characteristics of a product created by mental effort, as compared to the physical effort of product development (Levy and Weitz, 2007). In the case of apparel and textile firms, this would include the firm's ownership of copyrights, patents, trademarks and trade dress. Protection is also extended to the sensual differences of a garment because this intangible attribute is often the element that provides a given brand its distinction from the competition.

Apparel and textile products are protected in a variety of ways from unauthorized use of the intellectual property that makes a given product distinct from competition. Copyrights are protections 'granted by law for original works of authorship fixed in a tangible medium of expression' (What is a Copyright?, 2010). These laws provide exclusive ownership of characteristics such as the textile print pattern of Burberrys plaid. Patents provide the owner with exclusive use of an invention, which would include a new textile material such as Gore-Tex fabric. Trademark protection offers exclusive rights to 'a word, name, symbol or device which is used in trade with goods to indicate the source of the goods and to distinguish them from the goods of others' (What are Patents, Trademarks,

Servicemarks and Copyrights?, 2010). The Ralph Lauren Polo logo is protected by trademark laws. Another protection of intellectual property that is important for innovators of textile and apparel products is trade dress.

Trade dress protects 'any nonfunctional characteristic of a product or packages appearance and feel' (Gelb and Kristnamurthy, 2008: 36). A common and less-subtle example outside of the textile and apparel industries is the color blue used on Tiffany boxes. Often, an apparel item creates trade dress through innovative marketing, and is protected under the law to prevent a competitor from making a direct effort to confuse the buyer with an offering that a reasonable person could easily confuse for the original. In 2008, Adidas was awarded US$305 million based on damages because an imitator willfully attempted to confuse customers with a product that offered a similar look that consumers mistook for Adidas products. Protection of intellectual property is critically important to defend the unique market position that innovative products offer.

Innovation is not limited to the physical product, but can also involve any aspect of the market mix activities including promotion, physical distribution and pricing. Any new and exciting way that an apparel company can position either its brand or product has the potential to offer additional value to the firm's customers. In terms of types of innovation, the most common and easiest to implement is continuous innovation. This occurs when an existing product is enhanced so that it is viewed as an upgrade to the existing product, such as offering new colors or fabrics in a line. Continuous innovation requires little or no change in consumptive behavior on the part of the consumer. Contrast the characteristics of continuous innovation with the process of disruptive innovation.

Disruptive innovation challenges and often changes the paradigm of an industry. Zara, the low-cost provider of fashionable clothing, has introduced disruptive innovation throughout the firm's distribution systems. Unlike the typical industry lead time of six to nine months to bring a new design to market, Zara has significantly compressed the cycle time to approximately two to four weeks. Once the item is in the store, it will either be sold or removed within four weeks. This disruptive sourcing innovation has seen Zara's customers return to their retail locations up to 17 times a year as compared with the industry average of three visits per annum (King, 2010).

An important aspect of apparel and textile innovation is safety. When a consumer purchases clothing, they must be assured that the garment is safe in terms of materials and workmanship. This is of particular importance to parents when purchasing clothes for infants and children. Textile and apparel safety is addressed by both government mandates and through third-party testing. In 2008 the United States (US) enacted the Consumer Product Safety Improvement Act to further strengthen the country's powerful Consumer Product Safety Commission relative to third-party testing of children's products, including apparel (Seven Macro Trends in the Textiles and Apparel Industry, 2009). Third-party testing is very important in terms of developing standards for innovative smart apparel and textiles products.

One of the areas of the textile and apparel industry innovation that is experiencing the most rapid technological advances is smart fabrics and interactive textiles. State-of-the-art fabrics are able to monitor the vital signs of the wearer, provide global positioning system (GPS) location data on the individual's location and activate an alert if the wearer has become unconscious. Smart textiles are being developed in North America and Europe with input from textile firms, research universities and technology and communications companies. This broad range of competencies works together to provide third-party verification of the effectiveness and safety of the innovative designs (Mcgrane, 2008). The US is leading the way in terms of addressing third-party verification of the materials used in apparel and textile products.

Recent legislation in the US requires independent testing for lead and phthalate content in clothing. Lead, being a very toxic substance, is particularly harmful to young children and is found in paints and dyes used in apparel. Phthalates are used to add flexibility to plastics that are commonly used in children's apparel for applications including printing inks, coatings, adhesives and plastic objects found on clothing (Seven Macro Trends in the Textiles and Apparel Industry, 2009). One of the most respected third-party performance standards organizations is ASTM International.

ASTM International develops detailed specifications for the performance and safety parameters to which textile and apparel products must be tested. The testing parameters include wear characteristics, flame retardant performance and specifying the types of fibers to be used for particular apparel applications. Furthermore, the organization develops global industry standard terminology, which allows all testing organizations to provide universally understood results of their investigations.

2.3.1   Case study: Converse Chuck Taylor All Stars – from innovative to old fashioned to retro chic

The history of athletic shoes (also known as tennis shoes, sneakers, basketball shoes, as well as variety of other names) begins with an important manufacturing innovation developed in 1892. At that time, Goodyear was a manufacturer of rubber shoes called plimsolls, and the firm was a division of the US Rubber Company. Goodyear perfected the process of attaching rubber soles to a canvas

2.1 Converse logo.

upper, and thus created the first mass-marketed athletic shoes, which were called Keds. The name sneakers took hold because the advertising agency used by Keds, N.W. Ayer and Son, highlighted the benefit of rubber-soled athletic shoes as being quiet on all flooring surfaces. Keds, however, were perceived as a general athletic shoe, and were not positioned as enhancing the performance of an athlete in any specific sport (The History of Sneakers, 2010).

Marquis M. Converse identified an unserved market niche in the athletic footwear market. Although Keds sneakers were very popular, they were not positioned as a high-performance athletic shoe. In 1917 Mr Converse released the world's first high-performance basketball shoe called the Converse All Star. This was a major innovation in the athletic footwear market, and the All Star was an instant hit with participants in the rapidly emerging sport of basketball. The following year, Converse devised a marketing innovation that continues to be used by every maker of athletic shoes, the celebrity endorsement (The History of Converse All Stars, 2010).

In the early part of the twentieth century, basketball was a scholastic sport, played by school children and college athletics. The professional game was a minor sport at best. Mr Converse got one of the greatest high school All American basketball players, Charles H. 'Chuck' Taylor to wear his All Star performance basketball shoes. Taylor became one of the most famous basketball players in the US, and performed on barnstorming teams throughout the country. These teams played the game at the highest level of athleticism, and Taylor and his teammates all wore the Converse All Star. Converse contracted Chuck Taylor, in 1923, to be the spokesperson for the All Star brand and added his name to the product. Thus the legendary Chuck Taylor All Star was born (The History of Converse, 2010).

From the 1920s through to the 1950s, other athletic footwear companies were formed, but none would have a major impact on the marketplace until the mid-1970s. Adidas was formed in the 1920s when brothers Adi and Rudolph Dassler produced the first training shoe. Puma began production in 1948 with the first dedicated soccer shoe, while Nike began production in 1962, but the impact of these competitors was years away (The History of Sneakers, 2010).

During the late 1940s and lasting until the early 1970s, Chuck Taylor All Stars were not only the basketball shoe of choice for players at all levels of the sport, but they were adapted by teenagers as the casual shoe of a generation. Teens of the post-war era were in search of an identity and a uniform to express their style, and they selected blue jeans and Chuck Taylor All Stars as their unofficial dress code (The History of Sneakers, 2010).

Converse continued to innovate, and in 1958, offered their flagship product in colors other than black and white. Teens could now color coordinate their 'Chucks' (a commonly used name that referred to Converse Chuck Taylor All Stars) with their outfits or choose a color to express their individual style. There seemed no end in sight to the shoe's success. This began to change in the mid-1970s when Keds introduced Pro Keds, the first serious challenger to Converse's dominance as the basketball shoe of choice. This competitor was marketed as a higher priced, higher performing alternative to the old technology offered by Converse. Once the All Star brand became vulnerable, the door was open for specialized sneakers from Adidas, Nike, Puma, New Balance and others that offered high technology and specialized performance against the 60-year-old Converse (The History of Sneakers, 2010).

The 1980s saw Converse and the Chuck Taylor All Star fall on hard times, and the ownership of the company changed a number of times. By 2001, Converse's market share had dropped so low that the firm was forced to file for bankruptcy protection. The brand, however, was saved when it was purchased in 2003 by Nike (The History of the Converse All Star 'Chuck Taylor', 2010).

Using a blend of innovative marketing and the reissue of many of the classic Converse Chuck Taylor All Star designs and colors, the brand became a hit with a new generation of teens who viewed the product as retro chic. Although not a targeted market, baby boomers who wore Chucks in their teen years also became a sizeable market for the sneaker as a throwback to their youth. During the long history of the Chuck Taylor All Star brand, over 800 million pairs of shoes have been sold, and at the time of writing, the brand remains very popular with both the targeted youth market and the baby boomers that are, in some small way, recapturing their youth (Some Things About Converse, 2010).

2.4   Future trends in innovation

The proliferation of social networking sites is having a profound, and not yet fully understood, effect on the innovation processes within the apparel and textile industries. Technology is driving a complex paradigm shift in terms of how marketing organizations communicate with their target market, as well as how the target market communicates amongst themselves. Electronic commerce provides the apparel companies with the opportunity to track the online behavior of customers in terms of what is and is not being viewed, the time the average customer spends at the site, buying behavior (every click of the mouse should be tracked to provide the most detailed profile of each customer's online behavior) and trend analysis to lead the innovation effort to better meet the shifting demands of consumers (Speer, 2009). Social networking provides apparel firms with the vehicle to create buzz for their new and innovative products.

The impact of social networking sites on the future of the apparel and textile business cannot be underestimated. 'About 25% of the nearly 80 billion visitors who venture onto apparel and accessories sites are heavy social networkers' (Moroz, 2008: 3). Online apparel sales reached $2.7 billion in 2007, and that figure is expected to grow to $41.8 billion by 2012. Internet sales are heavily influenced by social networks, and innovative marketing is required to attract the attention of internet- and social networking-savvy consumers.

Retailers are setting up Facebook and Twitter sites to encourage site users to interact with the apparel offerings of the retailer, as well as other visitors on the social networking site. Sears, for example, has set up a Prom Premiere page that provides prom goers with the opportunity to share pictures of the apparel selected and to discuss the offerings with friends and family (Moroz, 2008). Another shift that is taking place in terms of buyer behavior in apparel is the loss of interest in runway shows.

Younger customers, who have been raised communicating on the Internet via social networking sites, are members of the immediate gratification generation. These people do not want to wait six months to see and buy the latest fashions. Furthermore, if the fashions offered by apparel companies and designers are not to their liking, they wish to design clothing that matches their personal style and tastes (Speer, 2009). Mass customization provides an innovative solution for the apparel companies to meet the demand for customized, consumer-designed apparel at a price point minimally above mass-produced products.

Mass customization provides apparel manufacturers with the opportunity to individualize apparel items that are semi-custom and are designed by the consumer using mass production methods of manufacturing. The adaptation of mass customization allows an apparel company to react to the rapidly changing tastes of consumers while providing the product at a competitive price. Furthermore, providing apparel using this technique shifts the innovation process, to varying degrees, from the apparel firm to the consumer. This also reduces lag time from shifts in consumer demand to product availability as the manufacturer can monitor the buying behavior and adapt accordingly (Frumkin et al., 2007).

2.5   Sources of further information and advice

One of the most important sources of information regarding performance testing and global apparel and textile industry standards is ASTM International (www. astm.org). This organization is one of the most influential third-party testing and standards organizations in the world. Test parameters are set for a wide range of apparel and textile characteristics that provide a universal understanding relative to the expectation of product performance and safety. ASTM International provides detailed testing procedures to ensure the safety of products in terms of important characteristics, such as flame retardancy and material safety, which are of particular importance to manufacturers of apparel for infants and children.

Another important global organization that is focused on both the apparel and textile industries is the International Textile and Apparel Association (ITAA) (www.itaaonline.org). In terms of the organization's mission, the ITAA 'promotes the discovery, dissemination, and application of knowledge and is a primary resource for its members in strengthening leadership and service to society' (About the ITAA, 2010). Furthermore, the organization supports scholarship and research in all aspects of the textile and apparel industries through grants, awards and fellowships, and holds an annual conference where the industry leaders discuss the latest innovations in all aspects of the industry. The ITAA is a great source of information and support of the textile and apparel innovator.

Good sources of data on global textile innovation can be found in trade journals that focus on the state-of-the-art textile technologies. *Textile World* (www. textileworld.com) is an excellent source of current information regarding virtually all aspects of the industry. Topics that are addressed in each issue include technical updates, news about government activities that relate to the industry, yarn, fabrics and data on dyeing, printing and finishing. This industry journal is an industry leader.

The *Journal of Textile and Apparel, Technology and Management* (www. bioresourcesjournal.com/index.php/JTATM/index) is a quarterly publication of the North Carolina State University and is written for researchers and innovators in the textile and apparel industries. One of the most important aspects of the journal is that it brings industry-leading innovators together in one forum to discuss the most current research in the fields.

There are also helpful online sources of information and one of the most valuable to the apparel innovator is WWD Online (www.wwd.com/fashion-news/?module=tn). This site offers one of the most authoritative sources of information about the latest innovation relative to apparel fashion, retailing practices, and apparel markets. Another useful online source of data is Just-Style. com (www.Just-Style.com), which offers data on the latest developments about the apparel and textile trades. Just-Style is an excellent source of news and extended feature articles about the most current activity in the industry.

2.6   References

About the ITAA (2010) The International Textile and Apparel Association. Retrieved on 5 July 2010 from http://www.itaaonline.org/www/default/index.cfm/about-itaa/.

ASTM Standards (2010) ASTM International. Retrieved on 1 July 2010 from http://www.astm.org/SEARCHTEST/search_json.htm?query=apparel&collection=all&searchType=full.

Bowonder, B., Dambal, A., Kumar, S. and Shirodkar, A. (2010) Innovation strategies for creating competitive advantage. *Research Technology Management*, 53(3).

Bulik, B. (2006) Betamax vs. VHS all over again? *Advertising Age*, 77(15).

Cooper, R. and Edgett, S. (2010) Developing a product innovation and technology strategy for your business. *Research Technology Management*, 53(3).

Gelb, B. and Krishnamurthy, P. (2008) Protect your product's look and feel from imitators. *Harvard Business Review*, 86(10).

Frumkin, S., Bradley, S. and Hedge, S. (2007) The challenges of mass customization on emerging markets. *Indian Retail Review*, 1(1).

How Innovation Really Works: It's not just about launching new products. It's about executing path-breaking models that are both disruptive and sustainable. A look at the innovators who changed the rules of the game. (May 2010) *Business Today*.

Igartua, J., Garrigós, J. and Hervas-Oliver, J. (2010) How innovation management techniques support an open innovation strategy. *Research Technology Management*, 53(3).

King, H. (2010) Learning from nature: the innovative invader. *Business Week Online*. Retrieved on 30 June 2010 from http://www.businessweek.com/innovate/content/mar2010/id20100328_962752.htm.

Levy, M. and Weitz, B. (2007) *Retailing Management*. New York: McGraw-Hill.

Mcgrane, S. (2008) Smarter clothes. *Time*, 172 (2).

Moroz, Y. (2008) Social networking evolves into venue to target shoppers. *Retailing Today*, 47(4).

Self, D., Bandow, D. and Schraeder, M. (2010) Fostering employee innovation: Leveraging your 'ground level' creative capital. *Development and Learning in Organizations*, 24(4).

Seven Macro Trends in the Textiles and Apparel Industry (2009) *Just-Style*, *11* (V).

Some Things About Converse (2010) *About Converse*. Converse, Incorporate web site. Retrieved on 28 June 2010 from http://www.converse.com/About/.

Speer, J. (2009) Little brother is watching you. *Apparel Magazine*, 51(2).

The History of Converse (2010) *Sneaker manufacturers*. Retrieved on 28 June 2010 from http://www.sneakerhead.com/manufacture-converse-p2.html.

The History of Converse All Stars (2010) *The History of Converse, Including the History of Chuck Taylor, the Man, Chuck Taylors the Sneakers and More*. Retrieved on 28 June 2010 from http://www.insidehoops.com/converse-history.shtml#ixzz0s9RZ2JXr.

The History of Sneakers (2010) *The History of Sneakers*. Retrieved on 28 June 2010 from http://www.sneakerhead.com/sneaker-history-p1.html.

The History of the Converse All Star 'Chuck Taylor' (2010) Retrieved on 28 June 2010 from http://www.chucksconnection.com/history1.html.

What is a Copyright? (2010) US Copyright Office. Retrieved on 30 June 2010 from http://www.copyright.gov/help/faq/faq-general.html.

What are Patents, Trademarks, Servicemarks, and Copyrights? (2010) US Patent Office. Retrieved on 30 June 2010 from http://www.uspto.gov/web/offices/pac/doc /general/whatis.htm.

2.7   Appendix: glossary

Intellectual property – Characteristics of a product or service created by mental effort (as compared to its physical characteristics).

Point of differentiation – The benefit(s) of a given product that separates it from the competitive offering.

Mass customization – Providing customers with the capability to semi-customize existing apparel platforms that are produced using standard mass-production techniques, while providing the product at a competitive price point.

Trade dress – The characteristics of a product that are directly related to the function of that product and creates a unique identity, separating the item from competitive offerings.

3

Textile product development and definition

M. STARBUCK, Ctext FTI Textile Consultant, Leicester, UK

Abstract: This chapter seeks to explain the stages that are necessary to turn an idea into a successful development. To further underline the importance of these stages and how people's changing needs affect market forces, a brief summary discusses some of the changes in textile yarns over the last 20 years. The discussion then moves forward to show the effect of these changes on the development of ladies lingerie, considers how to plan for change and provides a small overview of possible future developments.

Key words: exploration, evaluation test and analysis, commercialization, capitalization.

3.1 Introduction

This chapter discusses the processes that are required to achieve successful product development. The complexities and minefields are many, so it is useful to show how these areas work. First, what is development? A well-known definition would be: the process of turning an idea into a successful solution, often by solving a need, for example, the internal combustion engine that drives the world's transport systems. The stages of development are outlined in the following section.

3.1.1 The innovation process

Every good invention starts with an idea. Within textiles, missions are decided if it can be argued that there is a sustainable competitive advantage that will add actual value, reduce cost and/or create or increase market share. These inventions can be subdivided into low level inventions, with low risk, or high level inventions, with high risk. With both options the commitment to develop the innovation comes down to market knowledge, technical experience and gut feeling. To reduce any uncertainty we have 'research' (an organised and structured methodology, planned and focused, with limited scope and a systematic way of finding answers to questions) followed by 'development' (the application of the invention). The life cycle of these two events can be summarised as follows:

- *Exploration* This is ongoing, using Macro trends, consumer trends, academic research or new technologies.

34

- *Evaluation* This is the response to a question and the screening of new ideas, technologies or innovative routes, resulting in the development of a business case.
- *Test and analysis* At lab level this is the systematic testing of the idea at relatively low cost, resulting in proof of concept. At pilot level this is a mini bulk analysis of the idea, including consumer testing to give an assessment of the value of the invention.
- *Commercialisation* This is the commitment to full scale launch with all the product and financial risk attached.
- *Capitalization* This is the assessment of alternative options for application of the idea.

As a part of the innovation process, protection of intellectual property should be considered. A good and proven route is to use a lab book at the commencement of the invention, recording the idea, time and place with any witnesses present, then following with patent protection and the use of professional advisers.

In my career of working within the lingerie and seamless garment area for over 30 years, with personal innovations in fabrics and machine modifications all designed to give either cost or market advantage, the main criteria have always been to have a firm belief in the innovation, coupled with the technical knowledge to carry out the idea, followed by the enthusiasm to overcome doubt in times of difficulty. In this time my experience has been as development executive with a very successful fabric company with markets in Europe and North America, and then fabric innovator within a research and development facility that had responsibilities for numerous lingerie brands both in Europe and North America. The time in these two areas of development provided invaluable experience in working with fibre, yarn and machine manufacturers, all designed to give life to the next market trend. To highlight some examples of how innovation has developed in textiles, it maybe interesting to consider the development of fibre and yarn over the last 20 years.

3.2   Nylon to Tactel

In the UK in the late 1980s DuPont were experimenting with the use of various spinneret shapes in their extrusion of nylon fibre. One cross section gave a Diablo shape that had very distinctive light reflection properties. The fibre company felt that if this could be used in carpets, when the pile was cut, full use of the cross section would be shown. What developed was the 'Tactel Revolution' within the ladies lingerie market. Various fabric developments showed that the fibre, when used in a part orientated state (POY), had unique appearance, handle and drape with additional anti-static qualities, together with an added advantage in processing that meant that one griege fabric could be engineered to give various fabric weights, which proved to be very useful in a fast-moving lingerie market.

3.2.1 Microfibre

The Diablo experiment was so successful that it defined the fibre used in lingerie because the consumer wanted softness and drape with easy care. As the market inventor it gave major sales advantage to the fabric company that developed the concept. The fibre manufacturer with all of their overheads then needed a follow-on to this success, so again innovation with the fibre producer gave progression and microfibre yarns of 1.0 denier per filament (dpf) and finer were formed. Now differentiation was with softness and a matt appearance (with an additional treatment of titanium dioxide) against Diablo. The market wanted to follow the successes and a natural move forward was to improve handle after shine, matt and drape. Work was undertaken to search the world for a successor, found in the super peach skin handle that the Japanese had developed with the use of very fine polyester filaments as low as 0.3 dpf encased in polyvinyl alcohol (PVA) so that the yarn could be processed. The PVA was extracted as an after treatment on the fabric. This gave a superior handle to standard microfibre and also made it far more expensive, but the major issue was the effluent containing the waste PVA, which was unacceptable in the environment, and costly to reprocess. The end result of this was that peach skin handle was not developed within Europe.

3.2.2 Blends

The same fabric company felt that additional markets could be exploited if it was possible to innovate within the natural fibre area; this coincided with the spinners of natural fibre experiencing diminished sales in the lingerie area, and was partly solved with the successful blending of micro nylon or polyester with cotton and other natural fibres. This promoted easy care, soft handle, and marl effects with double dye techniques without major cost issues. The drape and cool handle of microfibre was not reproduced within the blend, but the comfort factor was superior, together with the market knowledge that people wanted to wear cotton next to the skin.

3.2.3 Regenerated fibres

To satisfy the market's need to have cellulose next to the skin, there was a resurgence of viscose, but the low wet modulus led to shrinkage and creasing problems. This in turn led to derivatives like Polynosic's and Modal becoming very popular. Currently, trade shows have confirmed that Modal in its micro form of 1 dpf is very successful and that the new Modal Air, launched by Lenzing in 2009, is a potential source of even softer fabrics with a cellulosic fibre of 0.8 dpf of regenerated cellulose microfibre, spun on the cotton system to give superior handle, drape and softness. It can be blended with other fibres like wool, polyester or cotton, dependent upon the end use. The market trend for ultra soft cellulosic fibres next to the skin is sure to continue.

3.3   Sustainability

Increase in the world's population has led to increased demand for natural raw materials. Demand for regenerated cellulose obtained from wood pulp has seen whole forests replanted every 20 years, to satisfy the need for cellulosic fibre. We see ever more diverse fibres becoming popular, such as Bamboo, which is sustainable within the growing area but is environmentally dangerous as it is spun on the 'wet spinning process', useing carbon disulphide (CS_2). A true revival is hemp, which is sustainable both in the diverse areas where it can be grown and also with the new enzyme treatments that allow the fibre to be much softer. When spun with cotton, as explained by 'Hemp Town' (Naturally Advanced Technologies), hemp will reduce dependence on cotton, which is ecologically unfriendly because of its need for large amounts of water and insecticides to maintain necessary fibre yields. The use of linen is also increasing, with its similar growing cycle to hemp, and silk is ever-popular as a luxury fibre. However, the truly sustainable fibre is wool; with its thermal and moisture wicking properties, it is the king of the natural fibre world. A problem is its lack of comfort next to the skin, unless used at 19 micron or finer, which has sourcing and cost issues. It is my belief that we will see various unusual blends with regenerated fibres becoming more popular.

Visits to various trade shows in Europe have demonstrated that the trend to combine bamboo with cotton, wool with silk, and wool with modal is being developed in the lingerie area, and linen and hemp blended with cotton will eventually become available in the underwear casual apparel markets. The real innovation would be to have an after treatment to wool that would maintain the properties whilst adding micro softness without shrinkage. This could then be used in both outerwear and really comfortable underwear, with all the thermal and cooling characteristics that are naturally built in, without the need to have other fibres blended with wool that would reduce its natural performance.

3.3.1   Recycled fibres

To a large extent today's fibres rely, directly or indirectly, on energy obtained from oil. The issue is that this resource is becoming depleted and we all have to conserve our natural resources. This has resulted in various forms of recycling. Natural fibres and regenerated cellulose have been reformed into multicoloured course yarns and then used in the inside of fleeces for some time. To satisfy market trends in the last five years polyester chips have been extracted from post-consumer plastic bottles from Europe and North America, sent to China in empty containers, sorted and then sent on to Japan to be turned into polyester chips for yarn production. This is excellent, except for the carbon footprint created in transporting these products around the world. Again, however, the market has responded and now at the time of writing two companies are offering pre- and post-consumer recycled yarns. Unifi in the USA are offering polyester branded as Repreve, a range of polyester yarns that use the company's own rejected fibre and waste water bottles, reconstituted into polymer chips. The other company is Nilit, an Israel-based manufacturer of fibre, that has used pre-consumer waste nylon from its own plants to create a new fibre, EcoCare, which uses over 50% less energy and significantly less water in its manufacture.

In talking to both Unifi and Nilit, the first stages of development into fabrics that can be used in lingerie are underway. In summary, it would be reasonable to suggest that the development of fibres, whilst complicated, will continue in response to market forces.

The lingerie market has also seen major changes over the last six years, with clear differences developing between high street ready-to-wear, brand and couture. The differences can be explained in terms of lingerie sets as follows:

- The speciality top end couture market starting at $250, such as La Perla, where quality of design, embellishments such as gold plated fastenings, exceptional fit, Levers lace, exclusive fibre blends of all types, and niche marketing maintain the image and exclusivity that the price demands.
- The own brand market from $30 to $80 offers personal service, fit defined to that brand, fabrics with fibre blends of nylon, silk, cotton, polyester and regenerated fibres. To maintain differentiation, the latest technical aids are used to improve comfort and fit with marketing to retain customer loyalty, which helps maintain the price point.
- In private label high street stores, selling between $10 and $40, where innovation is limited by price, attention to fit is maintained with fabrics in cotton, nylon, polyester and branded elastane with some regenerated fibres but on a small scale.
- At the lowest price area, below $10, which includes some private labels and the market stalls, where lower priced fabrics of nylon and unbranded elastane, polyester and less refined cotton are common, fit is dependent upon style and manufacturing base.

The last two categories, with garment prices of below $10 and up to $40, require ready-to-wear stock to turn up to every four weeks in order to maintain customer interest, whilst the more expensive brand and couture categories work to a six month season. Product development is, therefore, also focused in very different ways. Couture and brand markets, with prices of between $40 and $250 require the new innovative fibres, currently mixes of modal/milk, linen/cotton, hemp, silk and bamboo with expensive Levers lace or the new Jacquardtronic fine gauge lace. Garment shapes are refined but maintain the brand fit to ensure customer loyalty, with the latest technical modifications like titanium bra wires, gold-coated fastenings or innovations that give the wearer a feeling of exclusivity. It is quite common to expect product lead times of six months to two years in this area.

Ready-to-wear brands, with lower prices of below $10 and up to $40 have the opposite approach in reduced price, with product colour, motif, placement print, sewing thread colour or other areas that give the appearance of newness without adding cost or changing the technical performance of the fabric, which will change the fit of the garment. These areas are needed to maintain a short lead time with the additional complication of transport time.

In general the position today is that in ready-to-wear, brand and couture, garment design occurs in the country of sale, whilst manufacture follows the low cost sewing point. This is changing with the need for reduced lead times, where griege or dyed garments are transported from competitive cost manufacture, held ready for printing or motif application, labelled, packed and appear in-store. Alternatively some mid-priced garments are made in the country of sale but again with minimum changes. With all the changes, the complexity of creating, say, a ladies' bra with over 28 separate operations has led to the quality of the product becoming an issue, which is understandable when considering distance, culture and language. Therefore, it has been necessary to develop a system that confirms that what is designed is delivered and that what is designed can be manufactured. the following is one example of a workable system.

This system has been used successfully in a research and development centre supplying a multi-brand, multi-site manufacturing business. The business used Santoni knitting machines and developed garments for sale in the lingerie brand business with price points of up to $80 throughout Europe. The technology required that both fabric and garment were developed; this was due to the nature of Santoni technology where the fabric had to be developed on the same knitting machine as the garment was produced. The result was that two-stage innovation of fabric and garment was necessary to produce wearable products. Consequently any changes to the original yarn used in the development resulted in a change in the fit of the garment, so design samples, accurate specifications and reproducibility were vital. This process was unique compared with the other areas of lingerie, where the garment maker and the fabric supplier were separate stand-alone businesses.

The procedure identified a series of defined stages that were found to be essential to maintain design integrity, fit and timely delivery of the product, along with costing. Initially a Product Development Request was raised by the brand, which assumed that the fabric was acceptable. The critical dates for the garment were also included in this (i.e. cost target, brand, range sample dates, shipment date and launch to store). This triggered a critical path plan to encompass all the required actions, including the garment seals that ensured integrity and quality of the original sample. The sealing system was based on colour, purely because it overcame language barriers. It was as follows:

- White seal
  - design garment approval
  - fabric for handle, consumer benefits and performance
  - garment styling
  - wash and wear report
  - independent fabric test results
  - estimated cost.
- Green seal
  - base size of garment approval
  - sewing construction
  - component selection
  - wash and wear report
  - revised cost quote.
- Red seal
  - all sizes graded for fit and approved
  - wash and wear report for comfort and performance
  - estimated revised cost.
- Black seal
  - all sizes proven in production and approved
  - defined as correct against red seal
  - any variations agreed before final fit chart and specifications are circulated.

3.3.2 Salesman samples

These were required after the fabric and one colour had been approved, and they were then sent to the designated brand locations. When manufacturing the samples as a small bulk, this often gave an indicator of the garment issues to be clarified.

3.3.3 Colour development

The brand would decide which colours were required for that particular style, dependent upon country of sale and other garments within the range. The colour work was always done on the same base fabric as the white seal garment.

3.3.4 Planning

The complexities of the sealing system cannot be underestimated. It would only work with a proven critical path planning system, where all concerned would have their allocated tasks documented and agreed in a defined time frame. Accountability was the prerequisite for successful completion of the garment's product development.

3.4   Future trends

This chapter has discussed the complexities associated with the innovation cycle. It would not be complete without giving a brief personal overview of the market trend that may develop over the next 20 years. Globalization has been very successful due in part to the innovation of containerization. This seed change in manufacture and sales has led to the apparel textile market following low cost manufacture within the Pacific Rim. The economic wealth, education and standard of living are deservedly rising along with the need for innovation, which will take existing businesses out of textiles and into other profitable areas. There may follow another seed change that will return manufacture to the point of sale. This change would be to remove garment assembly, replacing it with complete garment technology. Yarns would be transported to suitable manufacturing locations within each country; they could then be manufactured and distributed in days, not weeks, with the reduction of energy, carbon footprint, usage, cost and time. The consumer would benefit from the speed of production and the irritation of a desired size being out of stock for up to six weeks would become a thing of the past. A possible starting point for this idea could be in the sports bra or control garment area, where a complete garment could be developed on one machine. Like most ideas, however, these can only be made possible through commitment to innovation.

3.5   Conclusion

All too often really good inventions are stifled either by lack of financial support or wilting enthusiasm. Both inhibitors are serious, but this author would encourage the reader with a burning idea to stay the course, with the knowledge that without innovation man would still be living in a cave.

3.6   Acknowledgement

My thanks to Mr Martin Bentham for his support.

3.7   References

McIntyre, J.E. and Daniels, P.N. (1995). *Textile Terms and Definitions* (10th Edn), Manchester: The Textile Institute.
Oxford Dictionaries (2010). *Oxford Dictionary of English*, Oxford: OUP.

3.8   Appendix: glossary

Complete garment technology – A term that describes garments made from yarn in one operation without the need for sewing or other ancillary operations.

Cotton system – A term applied to the manufacture of staple yarns on machinery originally designed to produce cotton (McIntyre and Daniels, 1995).

Diablo shape – Having a dog bone appearance.

Dpf – Denier per filament or the thickness of each fibre that make up the whole yarn.

Greige – Describes textiles produced before being bleached, dyed or finished, but that may contain dyed yarns (McIntyre and Daniels, 1995).

Jacquardtronic lace – Electronically controlled multi-guide bar warp knitting machines used to manufacture lace supplied by Karl Mayer, Germany.

Micron – Millionth of a metre (Oxford Dictionaries, 2010).

Partially oriented yarn (POY) – A continuous filament yarn with a substantial degree of molecular orientation, but with the possibility of further orientation (McIntyre and Daniels, 1995).

Polyvinyl alcohol (PVA) – A range of manufactured fibres composed of linear macro-molecules of polyethenol, with the propensity to be water soluble (McIntyre and Daniels, 1995).

Spinneret – Used to manufacture yarns, by a nozzle or plate with fine holes or slits through which a fibre-forming solution or melt is extruded to form fibres (McIntyre and Daniels, 1995).

Wet spinning process – Conversion of dissolved polymer into filaments by extrusion into a coagulating liquid (McIntyre and Daniels, 1995).

Part II
New product development of textiles

4

New product development in knitted textiles

S. EVANS-MIKELLIS, A.U.T. University, New Zealand

Abstract: This chapter focuses on the use of electronic flat bed knitting machinery. It discusses new developments in knitwear production and design capabilities, the quest to address issues of environmental responsibility through the use of seamless knitting technology and the potential for a change of approach to the design of knitted products.

Key words: seamless knitwear, Wholegarment, computer aided design (CAD) for knitwear, virtual textile design, digital printing for knitwear.

4.1 Introduction

The design of knitted products for the fashion market is a complex process, designers and manufacturers must respond rapidly to the changing needs of the consumer market. Since the 1850s improvements in knitting machine technology have transformed the speed, flexibility and productivity of the knitwear industry; however, from a design perspective it can be argued that until now, technology has in many ways prevented progress in knitwear. There have been advantages, like production speed and low cost products, but garment design has barely changed since the end of the nineteenth century. This chapter will investigate a new approach to design that is evolving as a result of the development of the most recent knitwear technologies with particular focus on seamless knitting, digital printing for knitwear and virtual knitting as a design tool.

4.2 Seamless knitwear

Until the mid-1990s the limitations of knitting machines meant that, with the exception of socks and gloves, there was no ability to manufacture knitted clothing. Knitting machines were only used to produce knitted fabric, and the fabric was then made up into garments (in the tradition of clothes made from woven cloth). This system is known as the 'cut, make and trim' (CMT) method and is still the convention for many knitwear companies today (see Fig. 4.1).

Alternatively, the fabric is knitted into flat, shaped pieces and then made up (see Fig. 4.2). This is referred to as 'fully fashioned' production, but the principle of the garment design is the same, and is based on flat pattern cutting. The only difference is that when clothes are made from woven cloth, features such as darts and ease are used to give the garments shape and fit, whereas in knitwear, the designer relies on the stretch properties of the fabric to provide garment fit.

45

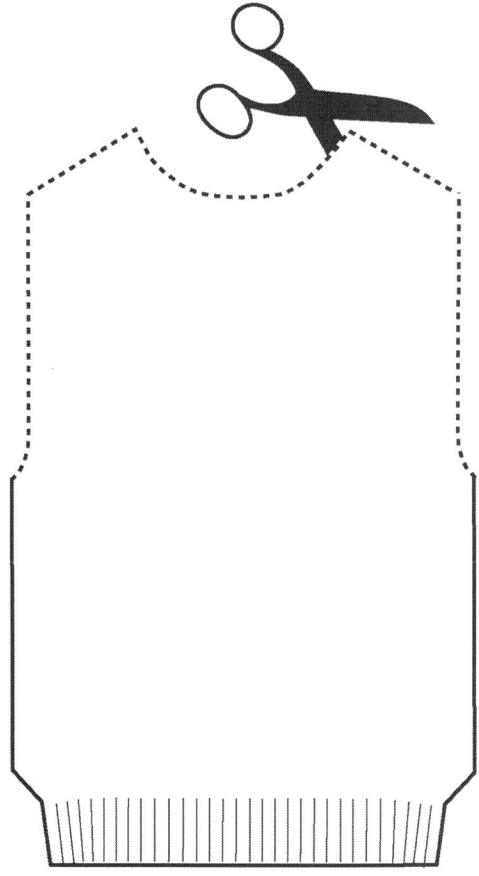

4.1 Fabric panel cut to shape.

The newest method of knitwear manufacture is that which produces seam free knitwear.

> At the 12th International Textile Machinery Exhibition (ITMA 95) held in Milan, Italy, Shima Seiki (the Japanese knitting machine company) made an historic presentation that dazzled the entire textile world. One-piece complete garment knitwear, the dream of the knitting industry, was finally a reality. Called WHOLEGARMENT® it signalled the dawn of a whole new era of modern knitting…
>
> (Shima Seiki, n.d.)

The Shima Seiki Wholegarment™ machine knits fabric in a series of tubes, which are interconnected (see Fig. 4.3). This makes it possible (for the first time) to create an entire garment at the same time as the fabric is created. These garments come off the machine complete and they have no seams. This method of knitting

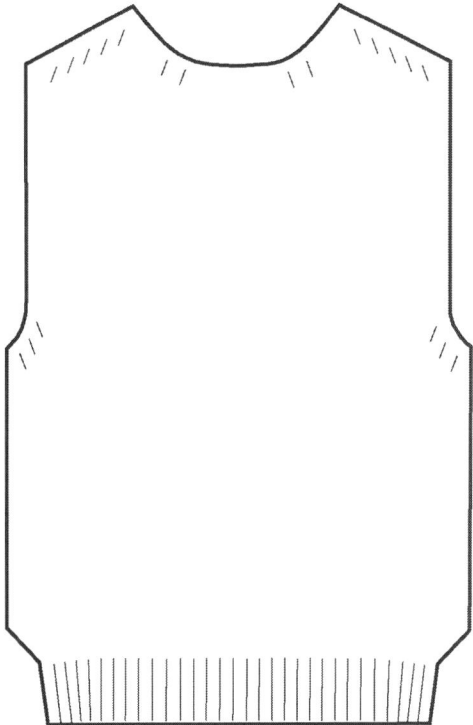

4.2 Fully fashioned fabric panel.

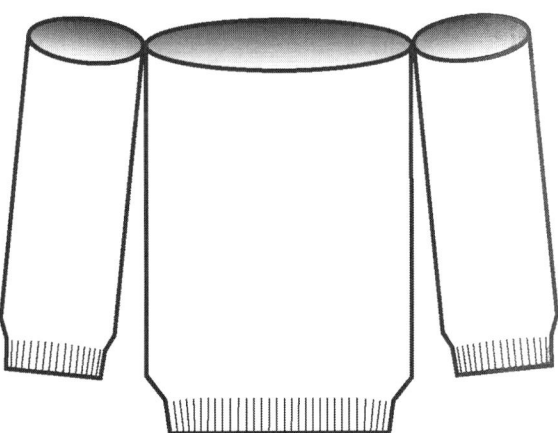

4.3 The tubular knit principle of the seamless Wholegarment™ sweater.

has the potential to achieve a superior product. It offers the consumer more comfortable knitwear with better fit as a result of eliminating the seams and it also eliminates all of the various cutting and sewing stages of garment manufacture, leading to savings in time and cost.

Seamless technology is becoming a successful method of manufacture. Hunter (2004) reports finding examples of seamless garments by companies such as Benetton, Max Mara, Burberry and Zara in European stores and confirms the use of seamless technology at both high and low price levels in the market. Many of the world's leading knitting machine manufacturers have now developed machines that have the ability to knit complete garments.

Although it is possible to knit tubular shapes very fast on conventional circular knitting machines, and some degree of seamless knitting can be achieved, the versatility of these machines is limited by the fact that it is not possible to do partial knitting (also known as flechage) within a course. Seamless knitting technology has this function and this is one of the features that make the design capabilities of the machines so exciting. It enables the designer to use a much wider variety of stitch structures than was previously possible and to create features such as integrally knitted trims.

On flat bed machines the obstacle to knitting seamless garments is the inability to knit a tubular rib with only two needle beds. Knitting machines with seamless technology use a method for manufacturing garments called the 'gore' technique. Also known as 'flechage', 'partial knitting' or 'short-row' knitting, this technique involves reducing (or increasing) the length of successive courses to give shape or to narrow (or widen) the fabric. Shima Seiki's Wholegarment system and Stoll's Knit&Wear machines both have the ability to combine many different stitch structures within one garment and they have the facility to knit different gauge areas in one course.

Seamless knitting machines can use intarsia carriers in conjunction with tubular knitting. Intarsia is a system of knitting that was originally used to create multi-coloured fabrics. Different sections of the fabric are knitted in different colours on separate feeders, because the pattern is designed around panels, or columns of colour, there is no yarn carried across the back of the fabric thus giving a true single bed, multi-coloured fabric. This process can be used on seamless knitting machines with the intarsia feeders being programmed to knit in specific areas instead of across the full width of the fabric. If different thicknesses (counts) of yarn are used instead of different colour yarns, a multi-gauge effect can be achieved and this can be used as an aesthetic design feature or as a functional element of the design, for example when areas of the garment are required to be less bulky (see Fig. 4.4).

Knitting machine manufacturers are continually refining and improving the production and design capabilities of their machines. The new technology is currently mainly used for garments that require the comfort of being seam free. The *Journal for Asia on Textile and Apparel* (ATA) has reported that companies

4.4 Knitted swatch using intarsia to create multi-gauge effect.

such as Reebok, Nike, Adidas and Calvin Klein are using seamless knitwear technology in the production of underwear and sportswear (Lam, 2005). British knitwear company John Smedley uses its website to promote the John Smedley ONE™ range, which consists of classic, simple knitwear pieces with an emphasis on seam free comfort (John Smedley, n.d.).

4.2.1   The benefits of seamless knitting

One advantage of seamless technology is the potential to cut manufacturing costs; no seams mean no sewing machinists, no construction machines or buildings to house them, etc. Yet there are also other benefits in terms of design, environmental responsibility and the potential for new market areas for knitted products.

In the Northern Hemisphere knitwear has always been an important component of both winter and summer garment ranges. In cooler climates knitwear is a staple part of the winter wardrobe and in summer the knitted sweater or cardigan replaces a heavy coat or jacket. In warmer countries though, there is always a challenge when selling knitwear for the summer season and in tropical areas, such as parts of Australia, knitwear can be perceived as too warm even for winter. To develop garments for this market requires rethinking traditional knitwear products and working with yarns and fabrics that are appropriate for the climate.

Apart from cotton T-shirting and interlock (used mainly for underwear), light, open, cool fabrics in fine yarns have sometimes been difficult to manufacture to a high standard of quality in fashion knitwear. This is due to the delicate nature of lightweight knitted fabrics that can ladder easily especially when they are constructed with smooth yarns in open stitch formations. These types of fabrics are often sewn using manufacturing processes such as overlocking, a method that is better suited for medium to heavy weight fabrics because seams can be bulky, or fraying can occur if the overlocking stitch is not tight enough. There is a higher level of skill involved in the production of lightweight knitted garments or knitwear made from yarns containing fibres with sheen, such as viscose and rayon. This also applies to other yarns that are popular choices for summer, such as yarns with slubs and knops and lightweight crepes and linens. However, now that there is the ability to make garments without seams, as Shima Seiki (2010) point out, new areas of the market, such as summer knitwear and eveningwear, are opening up for this type of knitted product.

Seamless knitting has given designers the possibility of using extra fine, lightweight and sheer fabrics without the need to accommodate overlocking or linking as methods of joining garment pieces. This is not only more satisfactory aesthetically, but it also improves the function of such garments, eliminating weak areas that are traditionally fragile due to the joining of panels, areas such as underneath the arms, or where trims like neck ribs are applied. The same thing applies when using fabrics that have very open stitch structures, loose mohair fabrics, ladder stitches, stretch fabrics and knitted lace can now be used in more inventive ways with fewer of the previous restrictions imposed by garment construction and concerns about durability.

Reversible knitwear is also possible when using seamless knitting technology, enabling designers to exploit double-faced fabrics, giving consumers garments with a choice of fabrication depending upon how they want to wear them. The ability to wear a garment inside out and upside down can also be utilised by clever designers to make garments that can be worn in different ways to create a variety of silhouettes.

4.2.2   Knitwear manufacturing and environmental responsibility

In the knitwear sector, as with other areas of the fashion and textile industry, the matter of environmental sustainability has become something that every designer should consider as an integral part of the design process. Knitwear designers now actively encourage the mindful production of fibres and yarns as well as having greater consideration for garment development and production methods that are less wasteful of resources. In 1992 the United Nations Industrial Development Organisation (UNIDO) and the Japanese Ministry of International Trade and Industry (MITI) sponsored a study to determine ways to promote energy conservation in the textile industry. Amongst others, the following conclusions were drawn from the study:

...when multi-line, small-volume production type high value-added goods are produced, energy consumption may increase rather than decrease with production rationalization, in contrast with mass-production type goods....

It is reasonable to consider that ultimately desired energy conservation promoting techniques will depend on the development and practical application of innovative technologies in each specialized technical field.

(UNIDO, 1992)

It could be argued that seamless production technology uses resources more efficiently because it does away with the need for extra machinery, such as overlockers and linkers, in the production of garments. The extra electricity required to power overlockers and linkers and to light and heat or cool the buildings that house the machines does add to the manufacturer's carbon footprint. However, this could be weighed against the aspect of social sustainability in situations where this technology is replacing existing CMT production and the subsequent loss of jobs as a result of a new production model. It should be noted though, that the perception of seamless knitting as undermining many traditional jobs in the industry could be viewed as a temporary perspective and one that will alter as the knitwear industry adapts to changes. The seamless knitting industry is certainly providing employment for many people. In the city of Yiwu in China there are a hundred and fifty seamless knitting companies, at the time of writing, with a total of nearly twenty thousand employees, producing over eighty million seamless garments a year. Almost all of these companies are producing warp knitted clothing (Anon., 2009).

In the UK some are taking a different approach to the use of the technology. Projects such as Considerate Design, a London College of Fashion research project led by Professor Sandy Black and Dr Penelope Watkins, was set up with the intention of challenging the traditional design process for knitwear (Considerate Design, n.d.). The objective is to combine the use of body scanning technology to gain precise body measurements, with computer aided design (CAD) knitwear design systems finally resulting in an individual, made to measure, seam free, knitted product that can be further customised to the buyer's personal request. The project is attempting to address the issue of sustainability by using materials efficiently to minimise waste, reduce overheads, such as labour costs for garment production, and produce one-off garments, which, due to the customisation of fit, will have superior comfort that will hopefully contribute to longer use of the product.

4.3   Printing on knitwear

Digital fabric printing for woven goods has been available for the consumer market since the late 1980s. As the technology has developed it has grown into a practical option for many areas of clothing manufacture. Designers are drawn to the flexibility of digital print, which allows them to bypass the use of silk screens or rollers for production, to alter designs easily with minimum expense and to produce short runs of fabric cheaply and quickly (Doshi, 2006).

It is now possible for both seamless and conventional knitwear to be embellished with the use of textile inkjet printing. Companies such as Brother produce machines that are primarily designed for printing dense fine gauge knits such as t-shirts. However, there are more complex printing systems, such as those produced by Shima Seiki, whose digital printers were originally developed for use with seamless knitting, allowing garments to be knitted in a single light tone to which colour could then be applied by printing. The height of the printing head of the Shima system is adjustable enabling not only flat fabrics, but also fabrics with three-dimensional textures and complete finished garments to be printed. The system is fitted with a charge-coupled device (CCD) camera, which is used to accurately position the print head so that each side of the garment can be printed and the edges blended. This means that the print design can wrap around the garment from front to back with no visible join or interruption to the pattern.

4.3.1   The advantages of digital printing on knitwear

Digital printing enables a designer to use a much greater quantity of colours in one image or repeat than traditional screen printing, and unlike transfer printing it does not alter the texture of the original fabric, making it ideal for knitwear because the stretch properties of the garment are not compromised. The clarity of the finished prints is excellent and a wide diversity of designs can be achieved, from motifs using simple colour separation to highly detailed photographic prints (see Fig 4.5 and 4.6).

The apparel market has become increasingly competitive. Variety, product differentiation and quick turnaround have become key factors in the survival of many businesses. The flexibility of digital printing technology allows samples and short runs to be produced easily without the set-up costs associated with screen printing, and it is, therefore, very well suited to the current trend for short production runs. The use of knitwear combined with print is also a way of creating a more cost effective product. A more limited colour range of yarns can be purchased giving the producer the advantage of economy of scale when buying raw materials. A much wider variety of colour options can be offered to the customer with very little extra expense incurred. A supplier can offer a quick response to requests for new colourways and styles, and the same garment can be printed in many different ways offering variety of product with only a limited amount of knit programming and technical development required.

Designs are created on a computer and are transferred directly to the printer allowing the fabric to be printed almost in the same way, as if the design were being printed onto paper. The process does, however, involve coating the fabric or garment with a pre-printing treatment; this will vary according to the dyestuff to

4.5 Digitally printed knitwear by Alysha Gover (image courtesy of Alysha Gover).

be printed and the fibre content of the fabric. There is also a post-printing process that involves steaming the fabric to enhance and fix the colour.

Digitally printed knitwear can be seen in all areas of the fashion market from high priced products, such as those offered by designer Alexander McQueen (2010) to lower priced garments from UK company All Saints. It is also evident in niche markets. Companies such as New Zealand's Untouched World, for example, use digital print to create pattern on knitted possum, a yarn that is not always suitable to use in jacquard structures due to its hollow fibre.

4.6 Digitally printed knitted dress (image courtesy of Alysha Gover).

4.4   Computer aided knitwear design (CAD) and virtual knitwear

Virtual knitwear is a digitally created image of a three-dimensional knitted fabric. It replicates the colours, the yarn type and the stitch structure to give the designers an accurate visualisation of a knitted fabric sample. The latest CAD systems, which enable the knitwear designer to produce virtual fabric swatches and garments, have improved the speed of the design process and greatly reduced the cost of sampling giving the designer more flexibility at this stage of product development.

Prior to the use of knit CAD, sample yarn would be purchased in multiple colours, designs would be graphed up and potential fabrics would then be fully programmed before test fabrics and garments could be knitted, often in several colourways. The use of the CAD systems enable the designer to create virtual swatches and garments in a limitless range of colours to present to a buyer before any yarn is purchased or any time has been spent programming machines and knitting samples.

The Shima Seiki Design System (SDS) is an example of one of the design packages that make it easy for the designer to turn concepts into virtual products. The system has features that allow the user to simulate fabric swatches, colourways and garments for use in design presentation, sales and manufacture. Figures 4.7–4.12 show the creation of a virtual swatch using this system. Machine gauge and single or double bed options are selected, and the SDS One automatically calculates the stitch and wale size to create an accurate graph (Fig. 4.7).

The loop simulation function then transforms the graph into a plain fabric for the user to work onto directly (Fig. 4.8). There is a stitch database that holds over 1000 stitch structures (Fig. 4.9) and this also contains all of the technical information required for programming the knitting machine (Fig. 4.10).

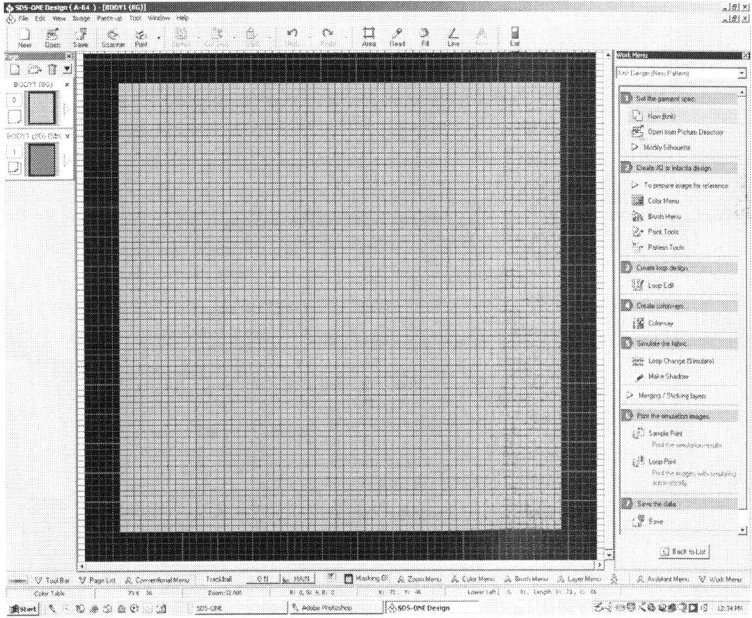

4.7 SDS One graph.

4.8 Loop stitch function.

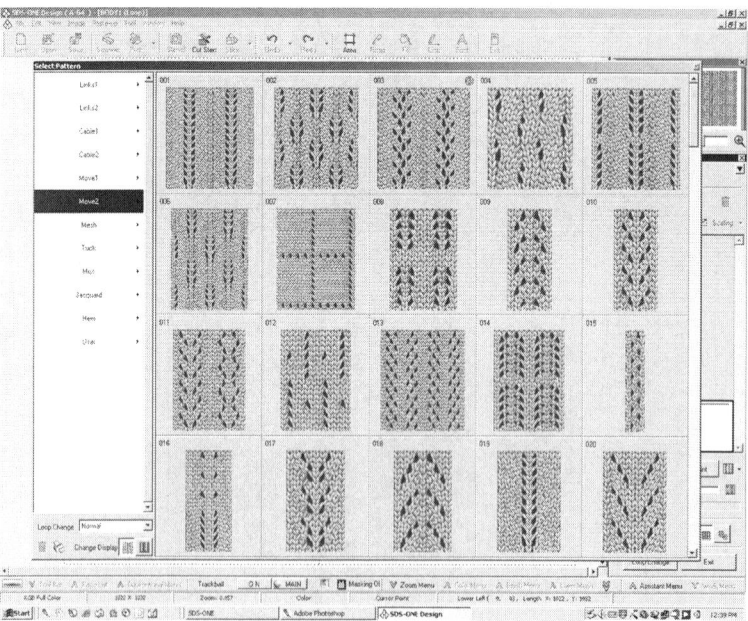

4.9 Stitch database.

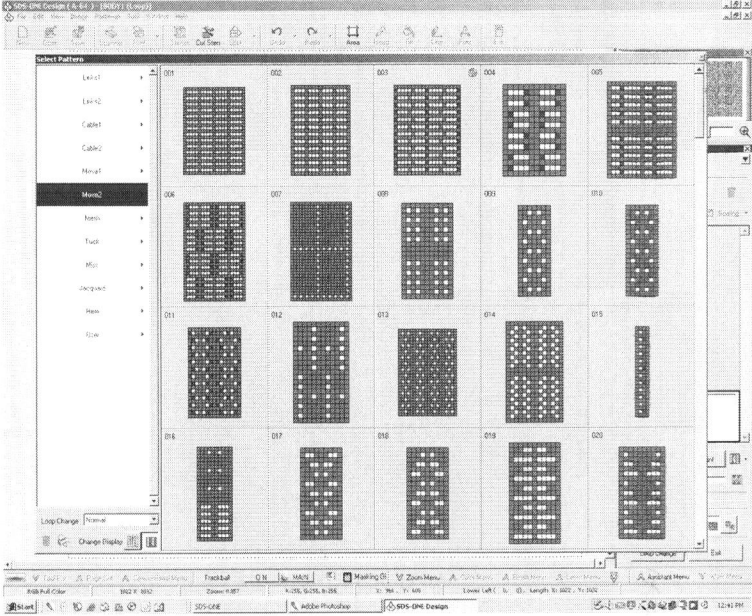

4.10 Programming information in stitch database.

Stitches can be selected and added to the base fabric and an infinite number of new fabrics can be created. This software enables knitwear designers of various skill levels to create designs in an efficient way. For experienced, creative knitwear designers who don't want to use what could be regarded as a 'cookie cutter' approach, stitch structures can also be created without the database. To do this, the designer selects a machine operation from the toolbar and draws the design onto the fabric (Fig. 4.11).

The SDS system also contains a yarn database that can be added to by scanning new yarns in. In this way it is possible to visualise a fabric in different yarns before the raw materials are purchased. Yarn is selected (yarn count and number of ends can be simulated) and a virtual swatch is created (Fig. 4.12).

Virtual fabric can be made in any colour required and it's easy to create colourways using the SDS because the system has a full Pantone® colour matching system menu, as well as colour meter and colour picker options for mixing custom shades. Many colourways can be viewed simultaneously (see Fig. 4.13) and once colourways have been chosen, there is the option to print multiple or single swatches. The virtual colourways can save sampling time and ordering extra yarn in the early stages of range development.

Visual representations of garments can also be created; designers can select from a database of garment blocks and each block can be modified extensively to create the garment shape required (see Fig. 4.14).

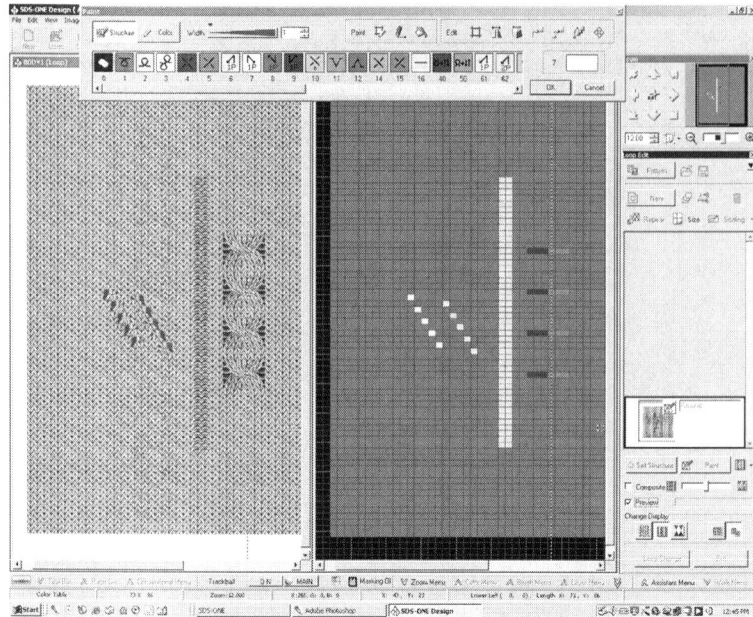

4.11 The split screen enables the designer to view both the vertical swatch and the technical information simultaneously.

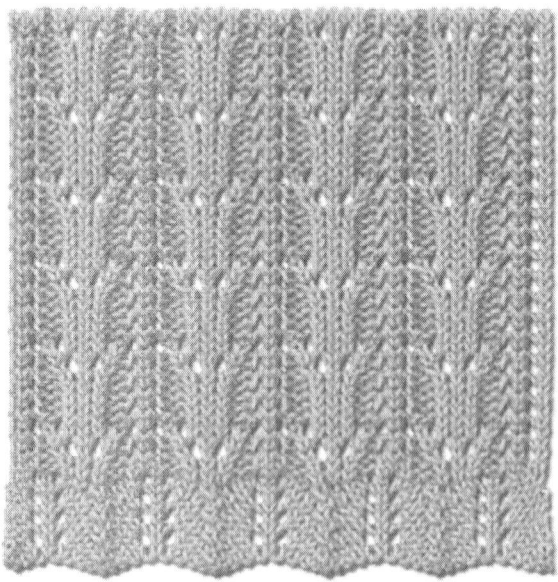

4.12 A virtual fabric swatch created using the SDS.

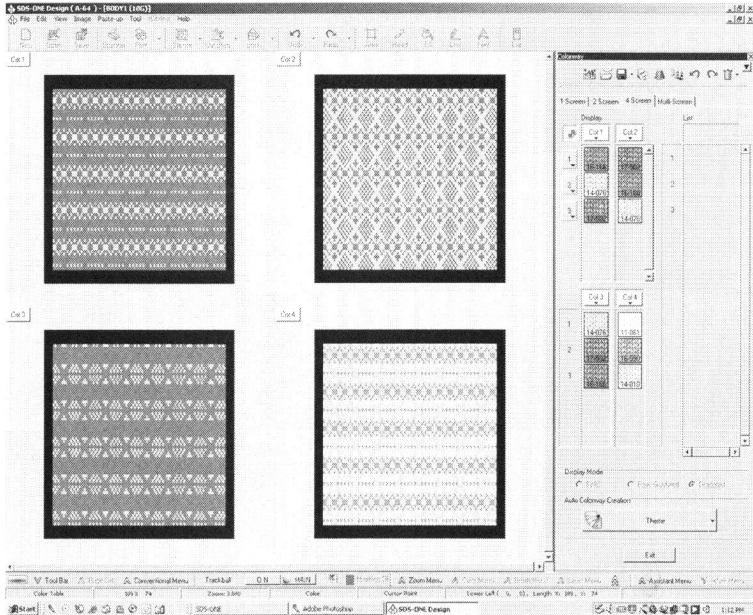

4.13 SDS screen showing fabric design in four colours.

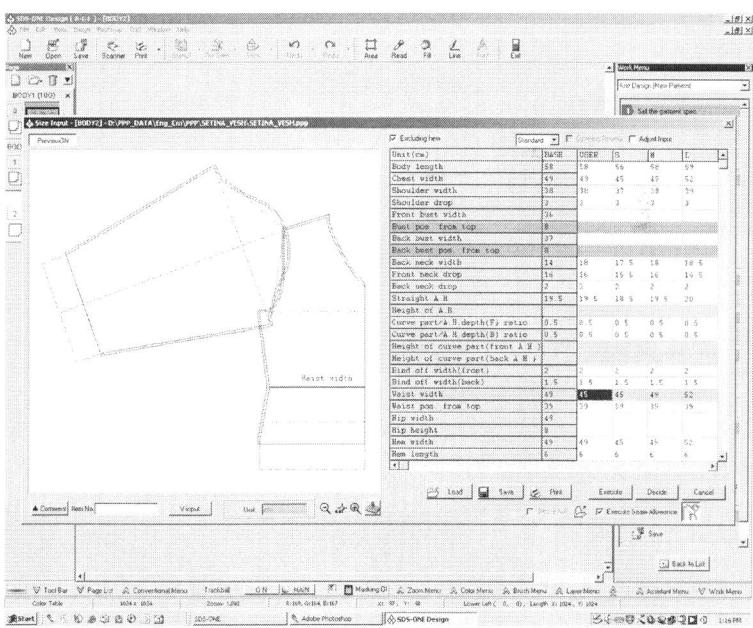

4.14 Pattern drafting using the SDS.

The fabric design can then be put onto the garment panels and the pattern can again be modified. When the design is complete, a representation of the garment can be printed (see Fig. 4.15).

Finished garments can also be presented on a mannequin using the template mapping feature. Simulated fabric is made first and then the designer can select the basic shape of the garment. Next a mesh is created to form a template. The fabric is then applied and the mesh is adjusted to follow the contours of the garment (see Fig. 4.16). Details of neck trims and ribs are put in last and the finished illustration could again be used for presentation of garments or colourways, without knitting extra samples.

Both these presentation methods are achieved by using the templates within the SDS system, but a designer wanting a more individual look can also use the mesh mapping techniques on their own photographs or sketches.

Other knitting machine manufacturers also provide similar technology. For example, the German company Stoll offers the M1 CAD system that utilises two computer screens, one that shows technical information in the form of programming data, and another that allows the designer to visualise the knitted structure (Stoll, n.d.). This way of designing and developing products ensures that all the yarn that is purchased by the manufacturer is pre-ordered by their customers, preventing yarn wastage and the shipping or freight of raw materials for sampling. It also reduces the need to send samples or to travel to customers, as all the virtual designs can be sent electronically. If the manufacturer has production bases in different

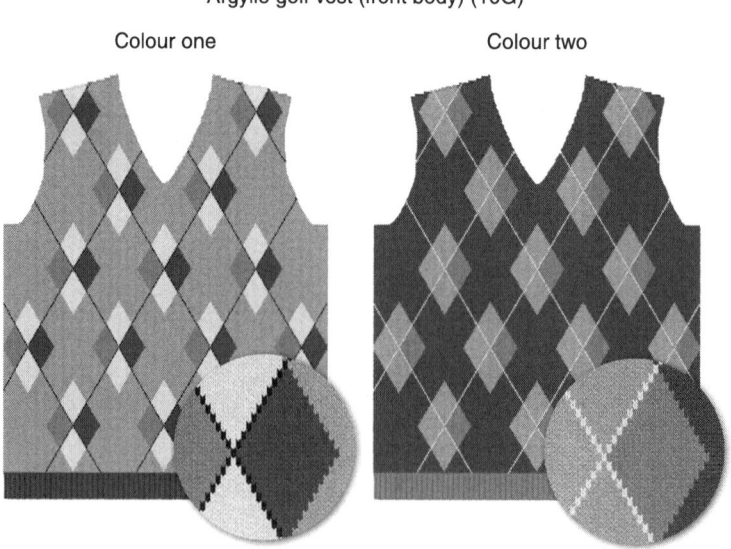

Argylle golf vest (front body) (10G)

Colour one   Colour two

4.15 A garment is illustrated along with fabric swatches shown to scale.

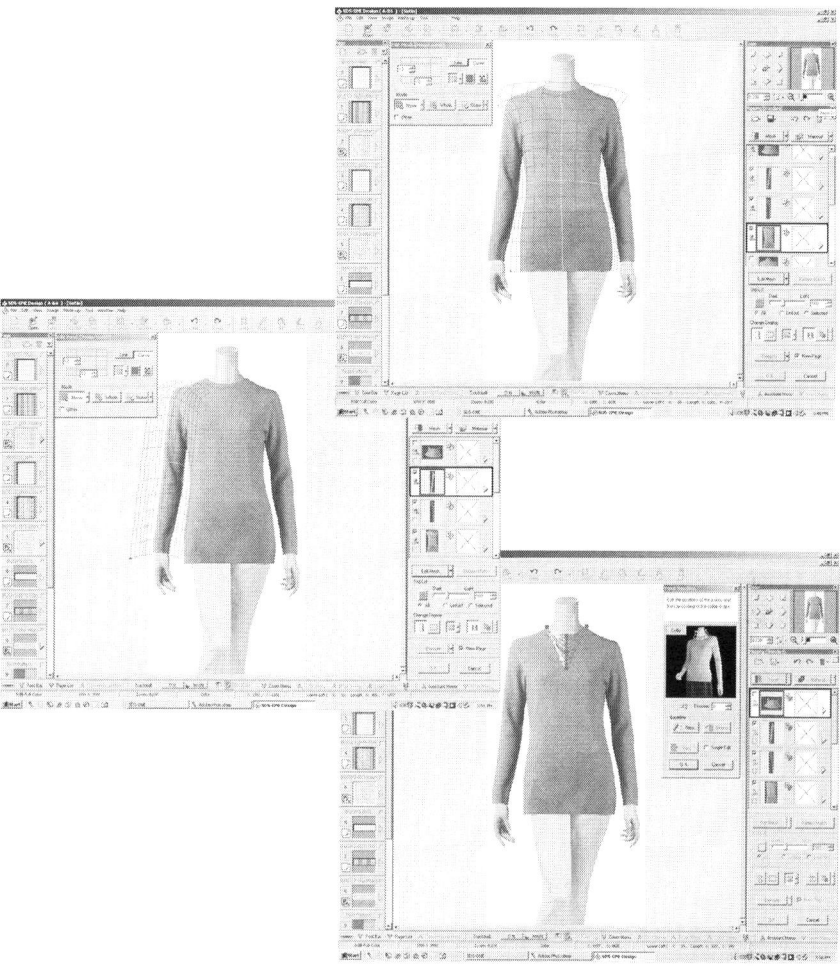

4.16 Template mapping using the SDS.

locations the final programming information can also be sent digitally once the styles have been selected. All of the information for each style is completely integrated and the consistency of communication provided is very efficient. It can be used for planning, designing, sampling, manufacturing and product promotion. Therefore, it is not only a tool for knitwear designers, but also yarn designers and spinners, knitwear technicians, buyers, merchandisers and retailers.

Because of the ease with which machine knitted garments can now be developed there is the potential for consumers to design their own individual, custom-made knitwear in the future. At the Sewn Products, Equipment and Suppliers of the Americas (SPESA) Expo in 2007, Shima Seiki presented this idea of distributing a product before it has been manufactured:

Just as newspapers are no longer manufactured in a central location and then distributed, Shima Seiki will show how garments might some day be delivered in digital form to your local drycleaners and then converted into the physical product to be worn.... Shima Seiki will exhibit its 3D knitting machine for whole garment knitting. Garments can be designed, viewed virtually, and shipped before ever being manufactured.

(Davis, 2007)

4.4.1   The implications for design education

In order to fully engage with the potential for a new direction in knitwear design we must consider the current training and education in this area. Presently design technicians are often trained within knitwear manufacturing companies and through courses offered by knitting machine manufacturers and further education organizations. This training has no creative or commercial fashion element with regard to the design of fashion product.

The people who are trained in fashion product design usually have very limited technical training, with knitwear often being an elective subject or a short paper forming a small part of a general fashion design course. Even education providers that are very well equipped to offer more in-depth training in technical aspects of knitwear often only offer students limited time to study in this area. The general view is that knitwear is a component of fashion design as opposed to a separate discipline. This approach perpetuates the CMT convention in knitwear design, where knitwear is treated merely as a stretch fabric from which garments will be made in the same way as woven garments are.

It is now time to address the changes in the training and working methods of knitwear designers that will need to take place in the wake of both the development of seamless production and CAD for knitwear to enable designers to utilise the full capacity of the technology.

A variety of academic papers have been written about the technological advances that have progressed the capabilities of seamless knitting functions, but little work has been done from a design practice perspective. When Shima Seiki first launched the Wholegarment machine it was hailed as the forerunner of an entirely new approach to knitwear design and production, however whilst this technology has been embraced from a manufacturing point of view, the design potential of seamless knitting has yet to be fully explored.

Seamless knitting technology has the capacity to integrate three-dimensional aspects of clothing design with textile development, as opposed to the usual practice of developing fabric and then making a garment from that fabric. Now that it is possible to build shape into what could previously only be produced as flat fabric, there is huge potential for a change in design thinking. There is an opportunity now for a new system of pattern cutting, or 'shape engineering' as it could more accurately be called, where the fabric and the required shape of that fabric are developed at the same time. This could also lead to other knitwear uses and products.

If we can break away from the woven CMT convention and fully explore integral knitting, there is great potential for commercial viability of new garment shapes enabled by this new technology. It requires a paradigm shift by those providing training as well as both manufacturers and designers to adopt this new mindset, to take us into a new era of three-dimensional knitted products.

4.5   Sources of further information and advice

Anderson, K. (2008, December). 'Seamless technology'. Retrieved February 2009 from http://www.techexchange.com/thelibrary/seamless.html.

Choi, W. and Powell, N. B. (2005, Spring). 'Three dimensional seamless garment knitting on V-bed flat knitting machines', *Journal of Textile and Apparel, Technology and Management*, 4(3): 1–33. Available at http://www.tx.ncsu.edu/jtatm/volume4issue3/articles/choi/choi_full_145_05.pdf [accessed 10 August 2011].

Eckert, C. and Stacey, M. (1994). 'CAD systems and the division of labour in knitwear design', *fibre2fashion*. Retrieved from http://www.fibre2fashion.com/industry-article/10/932/cad-systems-and-the-division1.asp.

Fibre2fashion (2006). 'Interview =N Shima Seiki Mfg., Ltd.' Retrieved 2 April, 2007 from http://www.fibre2fashion.com/face2face/shima-seiki/dr-masahiro-shima.asp.

Gupta, S. (2006, February). 'Mass customisation of seamless knitted garments', *Journal for Asia of Textile and Apparel*. Retrieved 6 June, 2009, from http://www.adsaleata.com/Publicity/ePub/lang-eng/article-31/asid-74/Article.aspx.

Hunter, B. (2004a). 'Complete garments – evolution or revolution? (Part 1)', *Knitting International*, 111(1319): 18–21.

Hunter, B. (2004b). 'Complete garments – evolution or revolution? (Part 2)', *Knitting International*, 111(1320): 22–23.

Hunter, B. (2004c). 'Complete garments – evolution or revolution? (Part 3)', *Knitting International*, 111(1321): 20–22.

Inteletex (2004, May). 'Shima offers new print option', Knitting International. Retrieved 5 October, 2008 from http://www.inteletex.com/FeatureDetail.asp?PubId=&NewsId=2936.

Mowbray, J. (2002, February). 'A quest for ultimate knitting', *Knitting International*, 109(1289): 22–24.

Peterson, J. and Ekwall, D. (2005). 'Production and business methods in the integral knitting supply chain', *Autex Research Journal*, 8(4). Retrieved December 2007 from http://www.autexrj.org/articles/37/254.

Power, J. (2007, Fall). 'Functional to fashionable: Knitwear's evolution throughout the last century and into the millennium', *Journal of Textile and Apparel, Technology and Management*, 5(4).

Spencer, D. (2001). *Knitting Technology: A comprehensive handbook and practical guide*. Woodhead Publishing: Cambridge.

4.6   References

Anon. (2009). 'Approaching Yiwu seamless knitting industry'. Retrieved 23 March, 2009 from http://www.2456.com/JasperWeb/Shows/Info/sid-239/nid-10821/lang-eng/MessageDetails.aspx.

Alexander McQueen (n.d.). 'Runway archive'. Retrieved 10 November, 2010 from http://www.alexandermcqueen.com/int/en/corporate/archive2010_ssm_womens.aspx.

Considerate Design (n.d.). 'Knit to fit'. Retrieved 9 July, 2009 from http://www.consideratedesign.com.

Davis, K. (2007). 'Shima Seiki and the "Cool Zone" at SPESA Expo 2007: hot technologies transforming the industry'. Retrieved 29 March, 2010from http://www.tcsquared.com/newsletter/2007/012407.html.

Doshi, G. (2006). 'The new wave of digital fabric printing technology'. Retrieved 2 November, 2010 from http://ezinearticles.com/?The-New-Wave-of-Digital-Fabric-Printing-Technology&id=372724.

Hunter, B. (2004). 'Complete garments – evolution or revolution? (Part 3)', *Knitting International*, 111(1321): 20–22.

John Smedley (n.d.). 'Cutting-edge technology'. Retrieved 19 October, 2010 from http://www.johnsmedley.com/cutting-edge-technology.

Lam, C. (2005). 'Seamless technology: trend and market'. Retrieved 12 January, 2006 from http://www.adsalepack.com/eng/epub/details.asp?epubiid=4&id=10957.

Shima Seiki (n.d.). 'Shima Seiki'. Retrieved 30 November, 2005 from http://www.shimaseiki.co.jp/wholegarment.html

Shima Seiki (2010). 'Fine gauge refinement'. Retrieved 1 November, 2010 from http://www.shimaseiki.com/wholegarment/.

Stoll (n.d.). 'Stoll-software solutions'. Retrieved 16 October, 2010 from http://www.stoll.com.

UNIDO (1992). 'Output of a seminar on energy conservation in textile industry'. Retrieved 12 December, 2009 from http://www.unido.org/fileadmin/import/userfiles/puffk/textile.pdf.

5
Fabrics and new product development

S. FRUMKIN and M. WEISS, Philadelphia University, USA

Abstract: This chapter discusses innovation in textiles and product development and the driving forces behind these advances. The chapter first reviews factors in the cultural, social and technological sectors that impact market demand. Textiles that respond to functional requirements, environmental sustainability demands and sensing issues (smart, interactive, electronic textiles) are reviewed. Lastly, textiles that marry artisan techniques with synthetic technologies are discussed.

Key words: nanotechnology, smart and interactive textiles, artisan techniques and collaboration, architectural tensioned membrane systems, environmental and sustainability considerations.

5.1   Introduction

In the field of textiles, innovation has resulted in fabrics that are lighter, smarter, multi-functional and with an array of engineered properties. These textiles retain familiar tactility, while providing consumer benefits ranging from moisture and temperature management to ultra-violet and antimicrobial protection. Through nanotechnology, fibers and yarns are gaining new functionality. Smart and interactive textiles respond to a variety of stimuli, including pressure, radiation and temperature, in both high-fashion and high-tech applications. Architectural tensioned membrane systems utilize textiles to manage light and temperature, as well as capture solar energy for building usage (Colchester, 2007).

Much of the innovation occurring in textile product development is a result of market demand. Consumers require specific functional attributes, as well as products that meet environmental sustainability considerations. Moving forward, textiles will become increasingly multi-functional. Advances in technology will enable the customization of products to meet market demands and the needs of an individual. Textiles produced using sustainable processes and practices will act as intuitive sensors and switches, creating a more comfortable environment for the consumer (Eufinger and De Schrijver, 2009).

Collaboration with innovators in diverse fields drives originality. Some of this innovation is occurring through the marriage of artisan techniques with synthetic technologies. Felting is combined with light emitting diodes (LEDs) creating new functionality. Stainless steel textiles are burner dyed. Phosphorescent fibers store and emit light. This development is fueled in part by the 'slow fashion' movement, with an emphasis on sustainable processes and

65

practices and a return to the familiar, tactile qualities that textile products exude in this technological era.

5.2   Market demand

Textile innovation responds to societal direction, cultural forces and market requirements. Specific functionalities may be imparted to textiles to satisfy the stipulation for ecologically smart products and processes. The global consumer is increasingly informed regarding environmental concerns and issues of sustainability. Consumers drive advances in this area with their purchasing power. As a result of technological developments, companies introduce to the marketplace products that solve challenges and enhance the consumer's environment, creating product demand. Properties, including liquid repellency, flame-resistance and shrinkage-resistance may be imparted through environmentally-friendly plasma treatments that alter the textile surface (Rathinamoorthy *et al.*, 2009).

Another approach to meeting consumers' environmental concerns is through the 'slow fashion' movement. Slow fashion involves purchasing products with a focus on organic materials, sustainable processes and fair trade practices. Designtex, a Steelcase company that was founded in 1961 with a focus on innovation, sustainability and performance, develops textiles primarily for the contract marketplace. As global citizens, Designtex worked with female weavers in Afghanistan to create the Common Threads Collection of area rugs. Designtex partnered with Arzu Rugs, a nonprofit, fair-trade organization that empowers women through education, access to healthcare and above-market compensation (Makovsky, 2009). Rug designs were created by Designtex and woven by weavers in rural Afghanistan. In the traditional Persian hand-knotted technique, these rugs were manufactured using locally produced, handspun, naturally dyed wool yarns (Singh *et al.*, 2009). This process that embodies the slow fashion movement responds to customers' demands for sustainable practices.

Textiles continue to provide a greater degree of flexibility in meeting a wide array of needs. The consumer determines the range of functions they desire from a given textile product, essentially customizing the product. Increasingly affordable manufacturing technology, married with market demand, has enabled this trend of mass customization, adapting mass production and processes to meet the needs of the individual consumer. In fashion, customers' expectations for comfort and fit have coupled with their desire for individuality in design, color and style. Mass customization allows the individual consumer to be involved in the process, opening a dialogue between the manufacturer and the customer.

Mass customization alters both the manufacturing process and the design of the product. Manufacturing is no longer producing one size fits all products. In designing the product, the company creates a basic structure or architecture on which the customer may build to meet their needs. Twenty-first century technology defines this era of customized clothing and accessories. In many cases, computer simulations are used to 'see' and approve the product prior to actual production. This product emphasis shifts the consumer focus from price-driven to feature-driven, thus increasing the perception of value (Frumkin *et al.*, 2007).

The textile product development cycle is a push-pull process. In some instances, demand for functionality drives advancement. This is the case with certain high-tech textiles developed for the outdoor/adventure market, where functionality is key. Textiles are created to meet very specific application requirements. For example, the sharkskin swimsuit famously worn by Michael Phelps of the US Olympic Team at the 2008 Beijing Olympics was engineered with a nanotechnology plasma coating to repel water molecules and reduce resistance (Eufinger and De Schrijver, 2009).

In other instances, technological developments, such as nanotechnology, allow the creation of new classes and applications of textiles. Nanofibers incorporated into ceramics produce flexible composites with lower breakage rates (Eufinger and De Schrijver, 2009). Textiles and fibers may be employed to reduce structural weight, creating high functionality with low density. Flexible, lighter weight products reduce transportation fuel consumption and increase product longevity. These are positive, sustainable directions responding to environmental concerns.

5.3   Functionality responses

In response to the market demand for functionality in textile products, advances are engineered at the molecular level through nanotechnology (Van Heeren, n.d.). Nanotechnology enables the creation of fibers with unique, inherent properties that in turn are the building blocks for textiles with a range of functionality. The resultant fabrics may be stronger, lighter, more precise and with multiple applications when compared to their traditional counterparts. Dr Karin Eufinger and Dr Isbel De Schrijver state, 'The future success of nanotechnology in textile applications lies in areas where new functionalities are combined into durable, multifunctional textile systems without compromising the inherent favorable textile properties, including processability, flexibility, washability and softness' (Eufinger and De Schrijver, 2009). Nanotechnology is employed in apparel to impart properties including easy-care (wrinkle resistance, stain repellency, odor prevention and quick-drying abilities), antimicrobial, insulating capabilities and ultra-violet light protection. Often, these technologies are developed for the high-performance or defense markets before finding applications to a broader audience.

Innovation in textiles and product development has resulted in significant growth in the 'shapewear' segment of the apparel market. Shapewear products utilize textiles that contain a percentage of elastane or latex for compression. Shapewear builds on the historic costume of waist cinchers and bustiers. This is the reinvention of a classic; capitalizing on current textile technology for varying compression levels, managing body temperature and moisture while incorporating comfort. Shapewear today includes products from undergarments to outerwear.

Fashion items exist with functional shaping built-in. Tank tops, crew necks, v-necks, long sleeve and short sleeve tops are all available in shapewear brands. According to Catherine Shannon, the Director for Design at Shapewear at Maidenform, Inc., this shapewear is 'meant to be seen'. New styles combine layering with shaping technology to 'feel good, look good – the emotional quotient of shapewear' (Shannon, 2010). Textiles for the shapewear category incorporate comfort, shaping, cooling and moisture management. Maidenform employs textiles that have long stretch and high modulus recovery, to create products that 'hug' rather than bind the wearer (see Fig. 5.1). Natural fibers are incorporated with new technologies whenever possible. Smart fabrics and nanotechnology are explored to create comfort with superior performance.

Additional developments in the shapewear market include the introduction of hypoallergenic foundation garments, in which the functional latex is encased in ribbed, plied cotton. The ability of cotton to wick moisture away from the body keeps the wearer comfortable, while at the same time providing a layer of protection from the encased latex (Torres, 2010a). Taking this category one step further, ShaToBu has incorporated strategically placed resistance bands within garments. According to independent tests, ShaToBu wearers burn 'up to 12% more calories' than they would when engaged in the same activity and not wearing the shapewear (Torres, 2010b). This aspect of textile innovation responds to both

5.1 Maidenform Flexees® Shapewear (courtesy of Maidenform).

the demand for functional shaping as well as the increased body mass of the aging global population.

Textile innovation and functionality are evidenced in a product developed to manage interior space. The Softwall, designed by Forsythe Mac Allen for Molo, is a flexible, expandable room divider made of polyethylene, unbleached kraft paper or tissue paper. Despite these lightweight materials, this modular freestanding wall system gains its strength from a honeycomb cellular structure. Softwall expands to hundreds of times the compressed size. These walls can divide a space in a myriad of formations. They act as sound insulators, create privacy and filter light. LEDs have been integrated into the structure to provide lighting opportunities. The flexibility of the LEDs lend well to the overall flexibility of the wall structure. The materials used to create Softwall are selected for their environmental sustainability. Polyethylene, tissue paper and kraft paper are all 100% recyclable and are manufactured with recycled content. Additionally, the adhesives used in the Softwall manufacturing process are non-toxic (Colchester, 2007).

5.4   Environmental sustainability responses

Given the increased market demand for environmental consideration, using 'green' technologies, reducing carbon footprints and employing sustainable practices, the textile industry continues to advance technologies toward this goal (Dharani *et al.*, 2010). Large internationally known brands and small boutique firms at both the manufacturing and retail levels are integrating sustainability as a core feature of their businesses. Raw materials in the supply chain, optimizing natural and renewable resources, energy and process inputs, transportation and distribution are all aspects under consideration.

In some instances, companies are making changes to their fiber selection or processing to produce a more sustainable product. In the *Industry Week* article 'Mapping global supply chain sustainability in the textile industry', it is noted that although the 'buzz around organic cotton will continue to increase', awareness is growing that organic cotton will be a small percentage of the total cotton market and of fiber consumption in general (Industry Week, 2010). From a sustainability perspective, the selection of natural agricultural fibers (e.g. cotton, linen, hemp, jute) is complicated by the frequent use of pesticides, chemical fertilizers or significant water consumption in the growing process. Alternative plant fibers exist that are not agriculturally cultivated, including raffia, aloe, abaca, nettle and kapok. The growth of these fibers generally does not involve pesticides. However, they exist in small quantities that do not permit mass consumption. Fibers such as these often show up in boutique textiles and products or as part of the slow fashion movement.

Synthetic fibers and fiber processing are evaluated relative to environmental impact. Polyester, for example, is manufactured from petrochemicals. Polyester will not decompose naturally. One manner in which this problem is tackled is through recycling. Teijin Fibers Limited has developed a closed-loop system, ECO CIRCLE®, to convert used articles of polyester clothing into new apparel items. In a closed-loop system, waste materials become raw materials for another product. Reuse of polyester fibers in this manner reduces energy usage and emissions (Nordstrom, 2010).

In response to environmental responsibility, consumers consider not only fiber selection, but also fiber processing. One answer to this challenge is the use of nanofibers, in which particle size may be altered to impact visual color. Morphotex® Structural Colored Fibers employ this principle. Also produced by Teijin Fibers Limited, Morphotex® fibers were developed using the biomimetic principle of the Morpho butterfly. The laminated structure is created through alternate layering of thin films of polyester and nylon, precisely controlling the layer thickness according to visible wavelengths (Kenkichi, 2005). The resultant fiber shares the vibrant blue of the Morpho butterfly. This cobalt blue color is a result of the structure of the fiber and is not imparted through dyes or pigments. Morphotex® fibers reduce energy consumption and environmental waste by eliminating dyeing in the manufacturing process.

5.4.1   Textile finishing

In textile finishing, transfer printing has traditionally been a heat-based finishing process on polyester fiber. As a more environmentally friendly alternative, the Cooltrans process has been developed to transfer print at room temperature onto a range of natural fibers. Using Cooltrans technology, fabric preparation, printing and fixation are all handled at room temperatures, similar to a cold batch process. This process reduces water consumption and energy consumption, as well as requiring less print paste than in traditional processes (Thiry, 2010).

5.4.2   Architecture

In the field of architecture, new textile applications are being developed to address the environmental concerns of energy costs, off-gassing of materials and employee wellness. Textiles bring flexibility and versatility to the skin of the building, creating an opportunity for the development of radically new structures. Consider a tensioned membrane system such as the Beijing Water Cube. The Water Cube, designed by PTW, Ove Arup and China State Construction Engineering Corporation's Shenzhen Design Institute, is made of ethylene tetrafluoroethylene (ETFE) membranes that form pillow-like structures and work in tandem with a steel frame construction. The ETFE membranes may be produced as transparent, semi-transparent or opaque surfaces, and may be printed with patterns that help control the light and patterning within the building. These ETFE 'pillows' expand when heated by sunlight, causing their patterns to overlap and controlling the sunlight entering the building. ETFE foils may be printed with photovoltaic plastics, creating the opportunity to capture solar energy (Colchester, 2007).

Structurflex has been creating architectural membrane systems for nearly 30 years, incorporating ETFE products, as well as polyvinyl chloride (PVC) and polytetrafluoroethylene (PTFE) textiles. One of the primary benefits to these systems is the design flexibility that fabric brings to the project. Forms may be developed that are aesthetically interesting and that blend or contrast with the site. In January 2010 artist Anish Kapoor, using an industrial textile to fulfill a vision, created an installation of tensioned membrane systems entitled The Farm for a private outdoor gallery near Auckland, New Zealand. Kapoor stated, 'I am interested in sculpture that manipulates the viewer into a specific relationship with both space and time' (Reitmaier, 2007). Kapoor created the sculpture to withstand environmental elements. It is formed by two identical steel ovals, one placed horizontally and one vertically, that support rich, red PVC coated polyester fabric stretched over monofilament cables (Fabric Architecture, 2010).

5.4.3 Designtex

In creating textiles for the contract marketplace, Designtex develops products that encompass function and fashion to enhance environments. According to Carol Derby, Designtex's Director of Environmental Strategy, 'Textile innovations have traditionally relied on novel materials or surprising constructions for their "wow" factor. What Designtex would define as innovative today are those same winning characteristics with the added serious backbone of heightened performance or increased sustainability, or ideally, both' (Derby, 2010).

One illustration of the innovative approach Designtex takes in response to environmental concerns is entitled Sonic Fabric. This textile, which is composed of 50% polyester and 50% recycled audio tape, was designed by artist Alyce Santoro. The fabric, used in a range of fashion applications, emits sounds when a tape head is drawn across the surface. It is woven by a small New England mill willing to work with unusual materials (Designtex, n.d.). The upcycling of the cassette tape and the artistic innovation of the fabric dovetail with the Designtex emphasis on sustainability and intelligent surfaces.

Sustainability is also evidenced in Designtex's project Bottles to Bags. This partnership with Rickshaw Bagworks creates messenger-style bags from textiles from the Designtex Regeneration Collection. These fabrics, created for upholstery use, are produced from 100% post-consumer recycled polyester. The polyester is developed from the PET resin created in the recycling of plastic bottles. Fabrics in the Regeneration Collection have earned Cradle to Cradle Silver Certification (Business Furniture, 2008).

The nonprofit institute GreenBlue, based in Charlottesville, Virginia, brings together industry experts to develop creative, practical solutions toward sustainable products and practices. GreenBlue, founded in 2002 by William McDonough and

Michael Braungart, is developing a Sustainable Textile Standard (STS). The STS will offer recommendations regarding all facets of the textile supply chain, from fiber selection, through textile processing, to product manufacturing and reuse or recycling opportunities.

To move toward a more sustainable future in textiles, True Textiles has developed Sustainability Waypoints as a guide. These ambitious goals include zero waste, zero harmful emissions, operating with renewable energy sources, designing closed loop processes and products, transforming transportation to maximize efficiencies, community engagement and the development of a new business model that encompasses economic, environmental and social impacts of the product development process (True Textiles, 2010). This approach will move the innovation and technological development of new textiles forward toward a sustainable future.

5.5   Sensing textiles responses

The marriage of technological developments with market demand has resulted in an explosion of new opportunities in textiles that respond to sensory input. Through nanotechnology a new class of textiles has been created known as quantum tunneling composites (QTC). QTC technology involves nanoparticles that are incorporated into fiber or yarn, so that the resultant textiles respond to a variety of stimuli including pressure, temperature and radiation. These textiles may be used as sensors or switches in apparel applications. Qio Systems applies QTC technology in their PANiQ line of integrated technology garments. These garments marry consumer electronics with traditional apparel applications. QTC textiles are employed by the National Aeronautics and Space Administration in combination with robotics for use on unmanned space missions (Wilson, 2010). The flexibility of textiles enhances the use of QTC sensors in prosthetics. The sensor may be programmed to respond to multiple stimuli at a desired level of sensitivity. This sensor, when combined with a textile that has tactile qualities similar to skin, creates an interesting area for future development (Wilson, 2010).

Technological advancement in functional sensing is present in interior textiles that employ photochromic and thermochromic dye technology (Periyasamy and Khanna, n.d.). Chromic colors are those that undergo a reversible color change in response to external stimuli. These color changes are a result of changes to the molecular state. Photochromic dyes are those colors that change through exposure to ultraviolet (UV) rays. When the UV radiation is removed, the dyes return to their original molecular state and the resultant original color. With thermochromic dyes, the visual color change is induced by heat. The ability to manipulate the critical temperature at which the color change occurs has a direct impact on usage. Applications exist for photochromic dyes in fashion accessories (certain sunglasses lenses), toys, cosmetics and garments. In the interior market, photochromic textiles could find application as window treatments or theatrical sets.

Thermochromic dye applications currently include home textiles and accessories, and novelty fashion apparel. If multiple thermochromic dyes are combined that have varied critical temperatures, a textile may be created that illustrates a range of color changes at different temperatures. This textile could be developed to respond to body temperature (as evidenced through emotional and psychological changes), human touch and ambient temperature. These interactive textiles require the user's involvement to create the visible color and pattern changes.

Recent advances in thermochromic technology have seen pigments incorporated into cellulosic fibers using the Lyocell process (Periyasamy and Khanna, n.d.). A textile woven of conductive, resistive and non-conductive fibers, printed with thermochromic colorant on the resistive fibers, can be electrically charged to change colors. Electrical power heats the textile, warming the thermochromic pigment, and effecting a color change. This process may be controlled through the electrical charge. These high-tech fibers are essentially disguised by the traditional aesthetic of the textile.

5.5.1   Smart textiles

Significant development is also occurring in the area of 'smart textiles'. Smart fabrics, alternatively known as interactive textiles, take on a variety of forms and are designed for a wide range of applications. Touch sensitive textiles are used to power bags and backpacks. Electronic textiles (e-textiles) may react to their surroundings, sense and respond to biometric feedback, provide an interface to personal electronic devices or alter the shape or visual aesthetic of a garment. Smart textiles may sense 'electrical, thermal, chemical, magnetic or other stimuli' (Van Heeren, n.d.).

In the couture market, e-textiles have been worn by celebrities on red carpets and seen on the catwalk. CuteCircuit, a London-based fashion firm, designed the Galaxy Dress featuring 20 000 full-color light-emitting diodes (LEDs). The LEDs are woven into sheer silk to retain a drapeable hand while allowing for a full spectrum of imagery and light. The garment, when not lit, is white in color. IPod batteries incorporated into the structure of the dress power the LEDs.

Wearable technology may be used to create a connection between people. CuteCircuit's Hug Shirt (see Fig. 5.2) allows two wearers located at a distance to connect remotely. One wearer 'hugs' himself, sending a signal to their friend through their cellular phone. The second wearer receives the 'hug' through vibrations in their Hug Shirt. For example, if the first person squeezed the right shoulder of their shirt, the second person would receive the signal as a vibration to their right shoulder. The signal is sent as quickly as a text message. The goal with these garments is to create an intuitive interface between the user and the garment; a garment that manages form and function seamlessly (CuteCircuit, n.d.).

5.2 CuteCircuit Hug Shirt (courtesy of CuteCircuit).

Another example of smart textiles is the Luminex® fashion line ranging from red carpet fashions to clubwear and sportswear. These intelligent textiles incorporate electronics that may be programmed to respond to environmental or human body stimuli. The fabrics are manufactured from a range of fibers, including polyester, optical fibers and nylon. Luminex® textiles are powered by rechargeable or traditional batteries, incorporated into the structure of a product or by a mobile phone battery charger through a designated port. Additional areas of exploration are concept car interiors and glass walls embedded with Luminex® textiles (Luminex, n.d.).

As technology continues to advance, textiles are a willing recipient. Textile technologies that at this moment are experimental or with limited availability will expand to meet market demand. Textiles that seamlessly incorporate optic or photo-luminous fibers will be made to react to light, heat and sound (Skorobogatiy, 2009). These textiles will enhance interactive opportunities between wearers/users and their environment. They retain the familiar, expected tactile and care properties, regardless of function. Additionally, the area of nonwoven technology will be explored for its aesthetic and performance opportunities, coupled with speed of manufacture. At present, nonwovens are relatively untapped in fashion applications.

5.6   Marrying artisan techniques with synthetic technologies

Much of the excitement in textile innovation is occurring through the somewhat unexpected marriage of artisan techniques and synthetic technologies. This is happening as a reaction to the simultaneous embracing and rejection of technology. As lives become ever more wired and connected, a desire for tactile, authentic experiences (the 'slow fashion' and DIY movements) emerges.

In fashion apparel, NUNO Corporation of Japan develops innovative textiles through the combination of artisan techniques with industrial technologies. Ancient traditions and the rich textile history of Japan are mined as starting points for new textiles. Synthetic fibers are combined with traditional aesthetics and centuries-old, creative processes (Hemmings, 2006).

The NUNO Studio explores textile development through art and science, tradition and experimentation. Disparate materials are combined and merged with unexpected finishing techniques. NUNO fabrics include burner-dyed stainless steel and phosphorescent fabrics with fibers that store sunlight then glow in the dark. Using industrial needlepunching technology, NUNO combines fabric scraps with raw wool to create new textiles. Shibori and origami serve as the inspiration for textiles that take full advantage of the thermoplastic properties of polyester for heat transfer printing and heat setting into colorful, pleated or puckered surfaces. The seeming dichotomy of industrial and artisanal processes when combined creates new textile expressions.

5.6.1   Felt

Another example of ancient processes combined with current technology is the widespread use of felt in contemporary applications. This nonwoven entanglement of fibers, typically in wool or a wool blend, is among the oldest known textiles, dating to between the seventh and second centuries BC, found in Pazyryk burial vaults in the Altai Mountains of Siberia (Brown *et al.*, 2009). In recent years, there has been a burgeoning interest in felt, as sustainability and the 'slow fashion' movement have embraced this traditional technique in contemporary applications.

Artist Claudy Jongstra's corporate felt installations are aesthetic solutions that control sound and light, define space, and create privacy. In her work for the Triodos Bank in Zeist, the Netherlands, Jongstra developed an installation for a glass conference room. The wool and silk felt, in combination with sheer silk panels, creates privacy while maintaining a desired degree of light and air. While beautiful, the felt artwork also serves important functions relative to the usability of the corporate space (Brown *et al.*, 2009).

Another interior, functional application is that of Cell LED (see Fig. 5.3). This felt carpet, designed by Yvonne Laurysen and Erik Mantel, is composed of laminated strips of wool felt that have been cut and pieced together on a flexible

5.3 Cell LED (courtesy of LAMA concept).

backing. In some areas, small wool felt nodes are inserted into the strips, creating a visual highlight. It is in these nodes that LED lights are placed between the strips and the backing, creating diffused lighting. One potential use for this carpet manufactured by LAMA Concept is to create directional, emergency lighting aboard aircraft (Brown *et al.*, 2009).

5.6.2   Maharam

Another approach to innovation comes from textile partnerships developed with industry outsiders. Maharam, a family-run business over a hundred years old, creates 'innovative textiles through the exploration of pattern, material and technique' (Maharam, n.d.). Primarily for the contract market, Maharam textiles are known for their high performance, married with innovation. Maharam embraces the knowledge and experience of leaders in various disciplines, while focusing on cultural and societal trends. These directions are synthesized with emerging technologies and textile engineering knowledge to create new lines, often incorporating unexpected materials.

An ongoing Maharam partnership is with Nike Sportswear. Nike, known for their innovation and designing toward performance and sustainability, has reinterpreted two of their famous shoes, the Nike Oregon Waffle and the Nike Blazer, using horsehair textiles. Nike x Maharam takes advantage of Maharam's extensive material knowledge, reinterpreting these famous silhouettes in luxurious textiles. Horsehair weaving is a centuries-old art. The creation of haircloth is labor-intensive, due to the inherent variation in the strands. Horsehair is a sustainable material; trimmed from a horse's tail, the hair then re-grows (Fiberarts, 2000). Patterns selected for the shoes are simple classics, checks, stripes and baskets; all rendered in rich black, allowing structure and luster to highlight the variations.

To create a new take on interior laminates, Maharam partnered with designer Luisa Cevese. Known for her proprietary process of embedding textile remnants – selvedges, end pieces, damaged cuts – in soft polyurethane, Cevese created 'Ply'. Cevese explores the marriage of the high-tech aesthetic of plastic with the soft, tactile qualities of textile fragments. Ply, engineered for contract upholstery, encases interlocking warp knit yarns in layers of polyurethane. Variations are created by changes to the types of yarn – chenille, tweed, cotton – and also to the opacity of the textile. The textiles vary from artisan to industrial in appearance, with soft mélange boucles giving way to hard twist, angular plies. When translucent, Ply offers visual layering opportunities, as a new consideration for upholstered furniture (see Fig. 5.4). Polyurethane, an organic polymer, offers the durability necessary for contract textiles.

5.4 Ply (courtesy of Maharam).

In Maharam's partnership with designer Hella Jongerius, new technology and artisan techniques were married to form 'Layers'. Layers was created through the combination of hand cut wool felt with full-width, mechanized embroidery. Wool felt provides a textural backdrop to the computerized needlework of machine embroidery.

Maharam also embraced artisan techniques in its partnership with Claudy Jongstra. Jongstra employs felt as her medium, but in contrast to the relative flatness of the industrial felt in Layers, Jongstra's Drenth Heath explodes in texture and tactility. Drenth Heath incorporates dimensional raw fleece into a wool/silk ground. The finished aesthetic is rough, wild and tactile, using technology to meet current performance standards, while evoking the ancient artisan technique.

5.7   Sources of further information and advice

Beylerian, G. M. and Dent, A. (2007). *Ultra Materials: How materials innovation is changing the world*, New York: Thames & Hudson.

Black, S. (2006). *Fashioning Fabrics: Contemporary textiles in fashion*, London: Black Dog Publishing.

Braddock Clarke, S. E. and O'Mahony, M. (2005). *Techno Textiles 2: Revolutionary fabrics for fashion and design*, New York: Thames & Hudson.

Cole, D. (2008). *Textiles Now*, London: Laurence King Publishing.

De Clerck, M. (2008). *Futuro Textiel 08: Surprising textiles, design and art*, Oostkamp: Stitchting Kunstboek bvba.

5.8   References

Brown, S., Dent, A., Martens, C. and McQuaid, M. (2009). *Fashioning felt*, New York: Cooper-Hewitt, National Design Museum, Smithsonian Institution.

Business Furniture (2008). 'Designtex products: green certifications', *Business Furniture LLC*. Available from www.businessfurnitureindy.com/pdf/green_certs_7=N08.pdf [Accessed 21 May 2010].

Colchester, C. (2007). *Textiles today: A global survey of trends and traditions*, New York: Thames & Hudson.

CuteCircuit (n.d.), 'Hug-shirt'. Available from http://70.32.91.33/products/thehugshirt [Accessed 12 May 2010].

Derby, C. (2010). Director of Environmental Strategy, Designtex, Interview by e-mail, 17 August 2010.

Designtex (n.d.), 'Sonic fabric'. Available from http://cdn.designtex.com//files/96eac830a 9fa495ab3a83501e42a2c6f/learn%20more%20about%20sonic%20fabric.pdf [Accessed 12 May 2010].

Dharani, E., Janani, S., Mythili, P., Balu, S., Surya, S. and Kumar, M. (2010). 'Green/eco-friendly textile manufacturing', *Colourage*, 57: 39.

Eufinger, K. and De Schrijver, I. (2009). 'Incorporation of nanotechnology in textile applications', Nanotechnology thought leaders series. Available from http://www.azonano.com/details.asp?ArticleId=2402 [Accessed 8 June 2010].

Fabric Architecture (2010). 'Anish Kapoor sculpture blends fabric and steel in New Zealand', *Industrial Fabrics Association International*. Available from http://fabricarchitecturemag.com/articles/0110_sk_sculpture.html [Accessed 8 July 2010].

Fiberarts (2000), 'Horsehair weaving revived', *Fiberarts*, 26(4): 26.

Frumkin, S., Bradley, S. and Hedge, S. (2007). 'The challenges of mass customization on emerging markets', *Indian Retail Review*, 1(1).

Hemmings, J. (2006). 'Exhibition review 2121: the textile vision of Reiko Sudo and NUNO', *Textile: The Journal of Cloth and Culture*, 4: 362–367.

Industry Week (2010). 'Mapping global supply chain sustainability in the textile industry', *Industry Week*. Available from http://www.industryweek.com/PrintArticle.aspx?ArticleID=21720 [Accessed 24 June 2010].

Kenkichi, N. (2005). 'Structurally colored fiber "Morphotex"'. Available from http://sciencelinks.jp/j-east/article/200513/000020051305A0526989.php [Accessed 23 May 2010].

Luminex (n.d.), 'Luminex news'. Available from http://www.luminex.it/pagine/news.html [Accessed 21 May 2010].

Maharam (n.d.), 'Company overview'. Available from http://www.maharam.com/news/company_overview.html [Accessed 21 May 2010].

Makovsky, P. (2009). 'Doing well by doing good', *Metropolis*, September: 76–77.

Nordstrom, G. (2010). 'Marine apparel, the second time', *Specialty Fabrics Review*, 95: 13.

Periyasamy, S. and Khanna, G. (n.d.). 'Thermochromic colors in textiles', Americos Industries India. Available from http://www.fibre2fashion.com/industry-article/printarticle.asp?article_id=804&page=3 [Accessed 23 May 2010].

Rathinamoorthy, R., Sumothi, M. and Jagadesh, S. (2009). 'Plasma technology for textile surface enhancement', *Textile Asia*, 40: 21–24.

Reitmaier, H. (2007). 'Anish Kapoor in conversation with Heidi Reitmaier', *Tate Magazine*. Available from http://www.tate.org.uk/magazine/issue1/descent.htm [Accessed 8 July 2010].

Shannon, C. (2010). Director of Design, Maidenform, Interview by telephone, 29 June 2010.

Singh, K., Wilk, D., McMenamin, M. and Chan, Y. (2009). 'Good looks, doing good', *Interior Design*, May. Available from http://www.interiordesign.net/article/478520-Good_Looks_Doing_Good.php.

Skorobogatiy, M. (2009). 'Bringing nanotechnology into fiber optics'. Available from http://www.photonics.phys.polymtl.ca/Overviews/Nano_into_Fibers_2009.pdf [Accessed 30 November 2010].

Thiry, M. (2010). 'Color it Green', *AATCC Review*, 10(3): 32–39.

Torres, A. (2010a). 'Flakisima develops cotton shapers', *Body Magazine*. Available from http://www.bodymagazine.us/news.php?idArticles=1548 [Accessed 6/23/2010].

Torres, A. (2010b). 'ShaToBu burns calories and shapes', *Body Magazine*. Available from http://www.bodymagazine.us/news.php?idArticles=1486 [Accessed 6/23/2010].

True Textiles (2010). 'Sustainability waypoints', *True Textiles*. Available from http://www.truetextiles.com/sustainability/sustainability_waypoints/ [Accessed 24 June 2010].

Van Heeren, H. (n.d.). Nanotechnology and lifestyle. Available from http://www.fibre2fashion.com/industry-article/printarticle.asp?article_id=1955&page=1 [Accessed 23 June 2010].

Wilson, A. (2010). 'Trends in technical textiles', *Specialty Fabrics Review*, 95: 60–63.

6

New product development in automotive upholstery

J. M. EASON, North Carolina State University, USA

Abstract: The automotive industry has played a vital role in industrial development since the early twentieth century and has become a driving force of manufacturing and globalization. Contributing to the success of vehicle brands has been the development of an industry whose sole purpose is to supply interior fabrics that meet the performance and aesthetic properties necessary for the automotive industry. As long as there are automobiles, textiles will be employed in their interiors. Understanding the new product development process for automotive textiles will increase the success rate of the vehicle and the industry.

Key words: automotive upholstery, automotive textile design, bodycloth, design trends, new product development.

6.1   Introduction

The automotive industry has played a vital role in industrial development since the early twentieth century and has become a driving force of manufacturing and globalization. Over the past five decades, the total number of cars and light trucks in the US alone has risen from 61.7 million in 1960 to 237.4 million by 2007, with annual sales figures peaking in 2000 at over 17.8 million units (Wards, 2009). Throughout the years, cars have come to represent American, European and East Asian culture, style, technology and affluence. Contributing to the success of vehicle brands has been the development of an industry whose sole purpose has been to supply interior fabrics that meet the performance and aesthetic properties necessary for the automotive industry.

This chapter focuses on new product development for automotive upholstery. In order to understand this process, it is necessary to become familiar with the automotive upholstery market, key drivers and supply chain. Also discussed in this chapter are novel materials and processes in automotive upholstery, along with future developments. Sources of further information and advice are included at the end of the chapter.

As long as there are automobiles, textiles will be employed in their interiors. Understanding the new product development (NPD) process for automotive textiles will increase the success rate of the vehicle and the industry.

80

6.2   The automotive textile market, key drivers and the supply chain

Both the automotive and textile industries have been pioneers in mass production. Weaving was the first industry to be fully mechanized, and the impact of Henry Ford's introduction of the assembly line is undeniable worldwide. Ford's efforts in the early 20th century revolutionized the industry through the standardization of one model (Ford Model T) at increasing volumes, which resulted in lower production costs, making personal transportation affordable and transforming people's lives. By 1966, however, a Yale University physicist determined that with Chevrolet's '46 models, 32 engines, 20 transmissions, 21 colors (plus 9 two-tone combinations), and more than 400 accessories', a customer could choose between more varieties of Chevrolet vehicles than the number of atoms in the universe (White, 1971: 189).

6.2.1 The automotive industry history and development

As the automotive industry developed, so did globalization and price pressures from foreign competition, which began getting the best of western automakers. Once among the biggest, most profitable and most glamorous of industries, the American automobile companies were no longer the industry's leaders and guiding light. Foreign competitors had emerged and pulled ahead in the eyes of their customers and the minds of the public, if not formally in industry statistics (Maynard, 2003).

Today, the automotive industry reflects the global economic crisis and US automotive sales in particular have experienced a drastic decrease, especially in the most recent years (dropping from over 17 million to 10.4 million in just five years), as seen in Figure 6.1. The American automotive manufacturers along with the Tier 1, 2 and 3 suppliers, who were once considered the international powerhouses of modern industry and innovation, are experiencing job losses, bankruptcy and the very real threat of permanent closure. While many factors have played a role in this turn of events, few were predictable or controllable. In attempting to survive, vehicles have become standardized and have lost the style and appeal they once mastered.

6.2.2 The automotive textile industry history and development

Automotive textiles can encompass a variety of products from upholstery, carpeting and headliner, to filters, tire cords and airbags (Powell, 2004). This chapter focuses on automotive upholstery (or bodycloth). These interior textile products have the most contact with the vehicle user and are outlined in Figure 6.2.

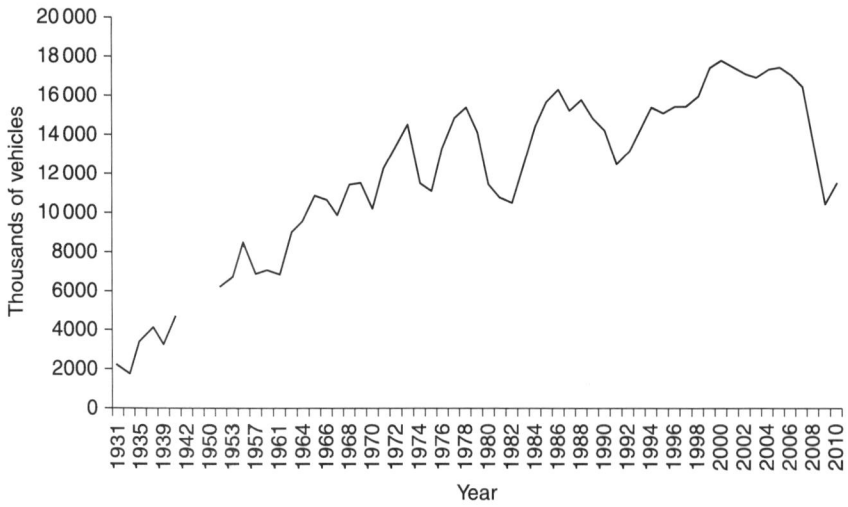

6.1 US vehicle sales, 1931–2010 (Ward's Automotive Group (2009), US vehicle sales, 1931–2008, retrieved from database).

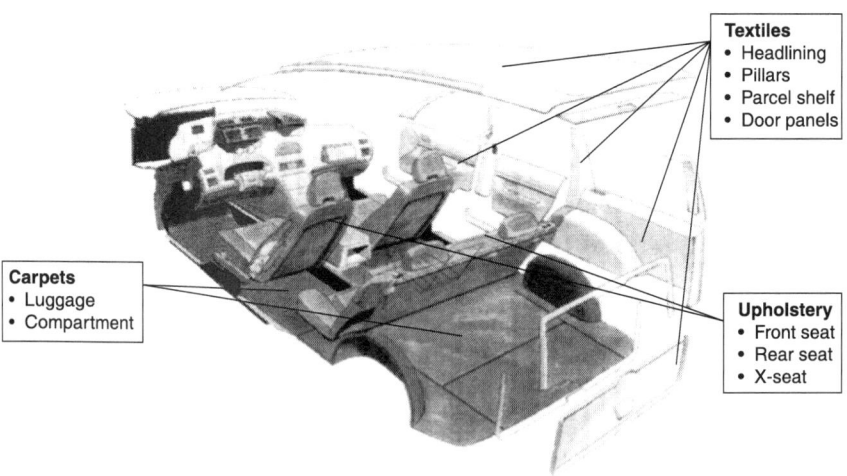

6.2 Automotive interior (Shishoo, 2008: 4).

In 2005 Powell researched the early evolution of automotive upholstery textiles in her paper, 'Transportation interior textiles: function and fashion'. 'The earliest vehicles had no interiors and only a wooden board seat which may have been covered with a scrap of leather, fabric remnant or horse blanket' (Powell, 2005: 2). According to Powell, as the country continued to industrialize both automotive and textiles, 'vehicles used a range of fabrics such as flat woven wool, cotton broadcloth, and mohair woven pile fabrics as reflected in the archives of suppliers such as Timmie Depoortere' (2005: 2).

Crucial to the evolution of automotive textiles, DuPont introduced nylon into industrial products during World War II. According to Mogahzy,

> with the quick industrialization of many innovative products to maximize the performance and safety of the armed forces, textiles would experience a major change in the 1940s. The synthetic fiber revolution had begun. The industry developed different forms of yarn from continuous spun yarns, flat to texturized, twisted to twistless, and plain to compound or fancy yarns. Numerous fabric types were developed within the three major categories of fabric, namely woven, knit, and nonwoven. Many specialty fabrics were also developed including crepe woven, dobby, pique, Jacquard, pile woven, double woven, braided and multiaxial woven structures.
>
> (Mogahzy, 2009: 4)

When textiles needed additional performance enhancements or modification of some form, 'the industry has always been ready to offer numerous types of chemical and mechanical finishing treatments or special coating and lamination' (Mogahzy, 2009: 4–5). In the 1950s, according to Powell, experimentation with new yarn technologies continued and was the period was dominated by 'flat woven fabrics with novelty yarns such as Lurex® and other bright yarn accents' (Powell, 2005: 5).

According to Colchester (1991), in the 1960s and 1970s textile manufacture was perhaps more affected than any other industry in the West by the idea of mass production. With the growth in supplier size, their technological capabilities also grew. Suppliers were expected to be able to offer all types of construction and secondary processes. 'Interiors began to resemble the tailored sofa of an upscale customer's living room. Fabrics would be button tufted, shirred, and gathered in an elegant, luxurious abundance reflecting the fashions of the era' (Powell, 2005: 7).

Powell also discussed the introduction of synthetic leathers or coated fabrics, such as vinyl, into the market.

> As the emphasis on durability of products became more important as more vehicles were purchased, used and resold, more and more coated fabrics were selected. If considering the geometry of the beach seat of the 1950s, it was possible to coat the backs of fabrics to improve performance such as seam strength, and wear, and contribute to flame retardancy. However, backcoatings made the product stiffer and less trimmable on a curved surface. Moreover, consumers experienced the heat and cold of the seat in vehicles with increasing amounts of glass and no or insufficient heating or air conditioning systems.
>
> (Powell, 2005: 6)

One solution to make vinyl seats more breathable and comfortable was to slit the vinyl into yarns for knitting and weaving (Powell, 2005).

By the late 1970s, according to Colchester (1991), the most industrialized areas (US, Europe and Japan) began sensing the danger of being undercut by rapidly developing countries (such as South Korea). The solution for the US, Europe and Japan was to focus on technical products that would be difficult for developing markets to imitate (Colchester, 1991). In addition to creating more technically advanced fabrics, increasing performance standards also helped ward off international competition.

6.2.3 Key drivers

Key design and engineering elements must be considered when developing automotive fabrics. These include standards and specifications, current materials and technologies, economy and sales, as well as consumer and global influences.

Standards and specifications

The automotive industry has some of the most stringent regulations of any global industry, which is reflected in the fabric requirements as well. While federal government regulations contributed to the addition of automotive textile safety products (such as seat belts and airbags), automotive upholstery standards were increased by original equipment manufacturers (OEMs) for safety, durability and sustainability. The list below includes the major necessary properties for automotive upholstery:

- abrasion resistance
- tear and tensile strength
- tensile elongation
- stretch and recovery
- seam strength
- dimensional stability
- snagging
- drape
- antistatic
- flammability
- comfort and breathability
- colorfastness and crocking
- lightfastness and UV degradation
- soiling and cleanability
- environmental ageing.

Another, more current focus of many automotive standards is on environmental issues. According to Shishoo (2008), the rate of reuse and recovery in Europe in 2006 was 85% by average weight per vehicle per year, while the goal for 2015 is

95%. Although many other markets do not yet have any requirements regarding the recycling of end-of-life vehicles, it is hopeful that they will follow suit with European regulations, which are already proving highly successful.

Current materials and technologies

Polyester is the dominant fiber in the industry due to its capability of meeting wear, fade and degradation, volume demand and cost pressures. According to Powell and Rodgers (2006: 3),

> the further development of manmade fibers provided the possibility of engineered performance to meet the demands of the automotive market ... for better durability, UV resistance and cleanability .... As these technological advances improved the function of the materials, changes in the design of seats and the passenger compartment would also expand the types of fabric formation used in automotive textiles.

Figure 6.3 demonstrates a traditional interior for vehicles from the 1960s and 1970s. The structure of these seats is relatively simple with little curvature. Figure 6.4 shows a traditional seat structure that is more common today. Both front seats are ordinarily the same structure and are separated by a center console. This sort of seat structure is known as a bucket seat, specifically contoured for one person. The primary fabric sections of a seat are the insert (A) and bolster (B). Bolsters are primarily lower cost, plain fabrics, while more expensive (and interesting) materials may be used on the inserts.

The impact that the evolving shape of seats has on upholstery is controversial and varies according to the market. Table 6.1 reflects the major technologies utilized in automotive bodycloth in the three key global markets in 2003.

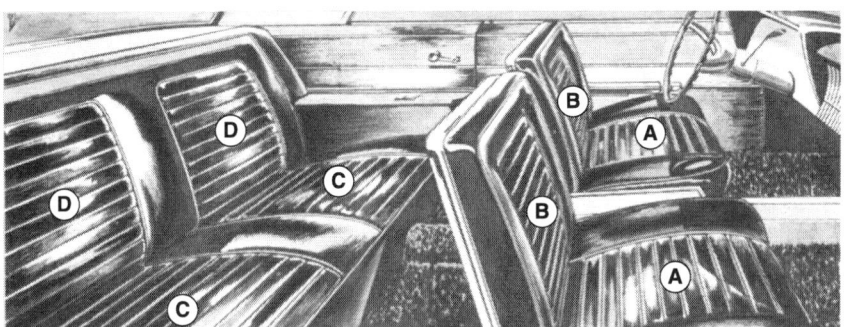

6.3 Interior diagram of traditional seating system (Detroit Body Products, 1964: 3).

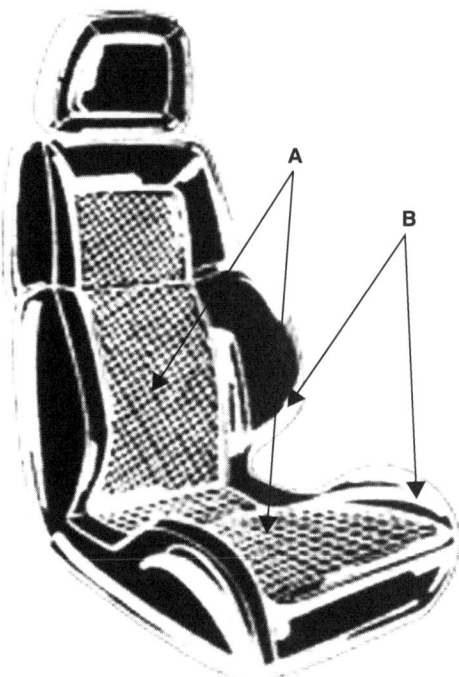

6.4 Traditional bucket seat (De Leo Textiles, 2004, cover).

Table 6.1 Automotive bodycloth construction by market

Fabric type	Global market usage (%)		
	Europe	USA	Asia
Flat woven	47	14	12
Woven velour	1	30	24
Tricot (including pile sinker)	15	11	44
Double needle bar raschel	5	23	9
Circular knitted	21	1	7
Leather	11	21	4
Total	100	100	100

Source: Anand 2003.

Economy and sales

All industries are driven by profit, particularly ones on as large a scale as the automotive textile industry. According to Mogahzy (2009), profit in the traditional textile industry has been primarily driven by the quantities it produces and the rate of production. The two main variable factors facing the textile industry are costs of the raw materials and labor costs (Mogahzy, 2009).

While raw materials contribute significantly to the overall manufacturing costs and are quite competitive worldwide, it is the cost of labor that sparked a major turning point in the textile and fiber industry in the early 1990s. The industry rapidly migrated from the USA and EU to Asia, Africa and South America, where general managers earn salaries that are 96.6% less than their western competitors (Mogahzy, 2009: 6). According to the Original Equipment Suppliers Association (OESA), 'the US has been a veritable dumping ground for goods that emerging economies produce but will not, or cannot, consume' (OESA, 2008: 115). This has resulted in a series of imbalances throughout the well-developed economies.

As global competition increases and the developed economies fall, the automotive textile industry is continuously cutting costs. In order to meet the cost-cutting demands of OEMs, suppliers are 'de-contenting' fabrics to create a lean base product for entry-level affordable transportation, similar to Ford's ideals of mass production.

> Not willing to compromise quality, OEMs expect higher and higher levels of performance standards and customer service in order to continuously improve the resulting products, while the pressures to reduce costs fall on the suppliers. In times of economic downturn, manufacturers challenge their suppliers to be innovative in finding ways to reduce costs without sacrificing quality.
>
> (Powell, 2004: 6).

Consumer and global influences

Another factor to greatly impact automotive upholstery is the role of the consumer and their connection to their vehicles. 'As the availability of affordable personal vehicles grew and the types of travel varied, so did the expectation on the interior. Beyond simple commuting, the majority of drivers attempt to socialize, eat, drink, conduct business, and entertain passengers in their cars' (Powell, 2005: 2).

The difficulty with the automotive industry is the long lag between design and delivery, which introduces a large element of risk into automobile manufacturing. 'Consumer tastes have to be predicted two to three years in advance' (White, 1971: 31). Understanding the factors that enter into a consumer's trim decision may help prevent marketers from wasting time and energy chasing the latest trend (Grossman and Wisenblit, 1999).

Maynard (2003) suggests that while emotional connection may get people interested in a vehicle, the grand majority of buyers are making their decisions based on practicalities. When it comes to interiors, consumers demand that appearance be maintained over the lifetime of its use or 'contribute to the resale value of a previously owned vehicle' (Powell, 2004: 6). In addition, the Internet and global communications have made consumers more knowledgeable about vehicles, options and financing before they ever enter a dealership. 'Rather than listen solely to Detroit, consumers now listen to each other. In an age of data, one of the most important criteria in buying a car is word of mouth' (Maynard, 2003: 30).

Additionally, demographics of the target customer are key to developing a successful product. OESA (2008) reports that for several years, the average vehicle buyer's age has been 46. 'Scion, the only brand on the market that has exclusively targeted young buyers, has succeeded as its average customer age of 38' (OESA, 2008: 88). In order to appeal to this younger demographic, Scion offers a wide variety of color, material and add-on options to provide the buyer with a 'personalized' vehicle.

An additional consumer and global factor, which majorly influences automotive upholstery design is fashion. While the European market desires large scale, exciting patterns, North American consumers prefer more conservative fabric designs (Powell, 2004: 11).

Perhaps the global influence most rapidly growing in importance is the increasing consciousness of the world's limits of dwindling fossil fuels and raw materials. 'There is now a growing awareness that we need to change the way that we have been living our lives' (Colchester, 2007: 22). The automotive and textiles industries are trying to do their part, some companies more quickly than others. 'Recently, in response to consumers' interest in environmentally responsible products, Toyota has announced 2015 as its goal for 95% of the materials in North American-made vehicles to be recycled', following suit with Europe (Johnson, 2005).

6.2.4   Supply chain and key players

In order to understand the NPD process for any industry, it is crucial to be familiar with the supply chain and key players. According to Powell, 'design and development are most successful when there is a collaborative partnership between the cross functional teams across the supply chain. The complementary or detractive nature of these interactions shape the process and the product in its priority and successful completion' (2006: 25).

Figure 6.5 represents the flow of the automotive upholstery supply chain and some of the key players in the North American industry. The automotive supply chain is organized by tiers from raw materials to the OEM, who receives a fully assembled seat, designed for a specific vehicle on the production line.

According to Powell, 'OEMs expect their global suppliers to supply any and all technologies specified at any assembly location whether in South America, Asia or Europe. An entire range of fabrics in construction, weights, widths, and colors must be available with the flexibility for rapid changes from one style to another if supply or performance issues should arise' (2004:3).

Powell and Rodgers developed a model of the automotive bodycloth selection process, seen in Figure 6.6. The process is initiated with OEM Reverse Presentations, which according to Powell (see page 91),

Tier 4		Tier 3		Tier 2	Tier 1	OEM
Fiber & chemical	Yarn/Hides	Upholstery materials (fabric, leather, vinyl)	Foam & laminate	Cut & sew	Complete seat systems & trim cover	OEM
DuPont	Unifi	Guilford Mills	Foamex	TechnoTrim	Johnson Controls	General Motors
Reliance	Reliance	Sage	Shawmut	HFI	Lear	Ford
Teijin	Teijin	Kawashima	Faurecia	CNI	Faurecia	Chrysler
Milliken	Performance Fibers	Seiren/Viscotec	Grupo Copo	Seiren/Viscotec	Magna	Toyota
	Aunde	Aunde	Woodbridge	Irvin	Aunde	Honda
		American Thierry		Tachi-S	TS tech USA	Nissan
BASF	Toray	Toray				Mazda
		Eagle Ottawa				Hyundai/Kia
		Seton				Volkswagen
		Garden State				BMW
						Mercedes
						Mitsubishi
						Isuzu
						Suzuki
						Subaru

Aftermarket

Dealers

6.5 Key players in the automotive upholstery supply chain, North America (Eason, 2009: 28).

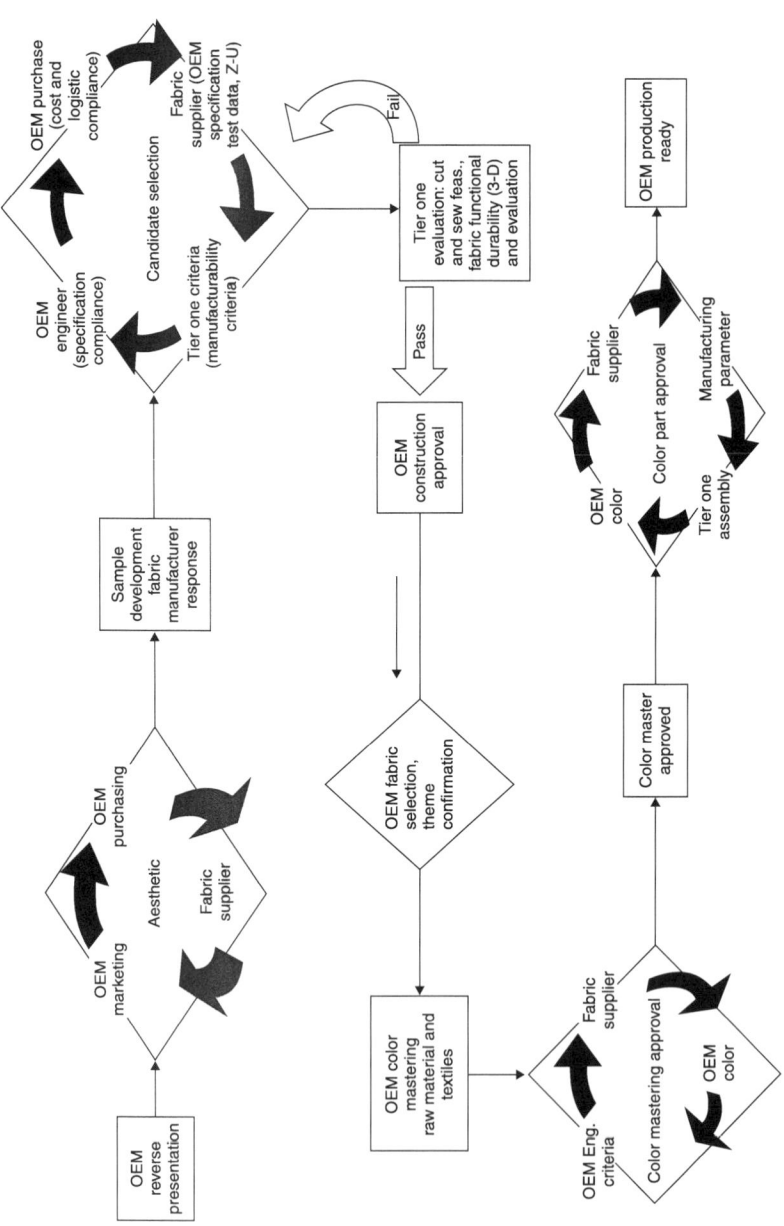

6.6 Automotive bodycloth selection flowchart (Powell and Rodgers, 2006).

are given by their color and trim team to selected suppliers usually including certified seat, leather and fabric suppliers. This is the official "kick off" of the new or freshening of an existing vehicle development. Communication of the brand image or new direction for the vehicle is enhanced with storyboards, inspirational images, market data, consumer socio-demographic profiles, color direction, and sometimes fabric swatches from other markets. A specific technology or construction may be specified at this time such as knitted pile or flat woven. The engineering and price parameters are established as critical parts of the total design brief. The seat design and manufacturer may or may not be disclosed at this point.

(2004:8)

As shown by the complexity of the supply chain and bodycloth selection flow chart, a lot of people can 'say no' along the way. For example,

once the fabric sample is visually approved by design then the certification process of the proposed fabrics begins to confirm quality and refine costs. Sample or "trim" yardage must be prepared and tested on the proposed seat designs. Feasibility studies and quality reviews must be completed before scaling up to production volumes (estimated at 68 million yards with the peak in automotive sales in 2000). Feedback may be given to the component supplier at any point with requests for changes in the product. Flexibility and responsiveness at each stage maximizes the opportunities for a total systems cost reduction.

(Powell, 2004: 5)

6.3   New product development process for automotive upholstery

Maintaining a consistent awareness of the new product development process is essential for managers as well as designers to generate successful ideas that have the potential of becoming successful products. According to Urban and Hauser, good proactive new product development processes must reduce risks and encourage creativity (Urban and Hauser, 1993). Urban and Hauser's five-step NPD process, known as the Strategic Plan, consists of opportunity identification, design, testing, introduction and life cycle management. This process is outlined in Figure 6.7. The Strategic Plan provides the background for developing and managing a product, such as automotive upholstery fabric. This process is useful to manage risk, which can lead to savings in expected time and cost, and identify a highly profitable new product (Urban and Hauser, 1993).

Unfortunately, new product development is rarely as straightforward as the model presented above. As noted in Section 6.2.4, automotive fabrics must go through many channels and approval stages at many different levels, before they are employed in the vehicle. This means that a lot of people have decision-making

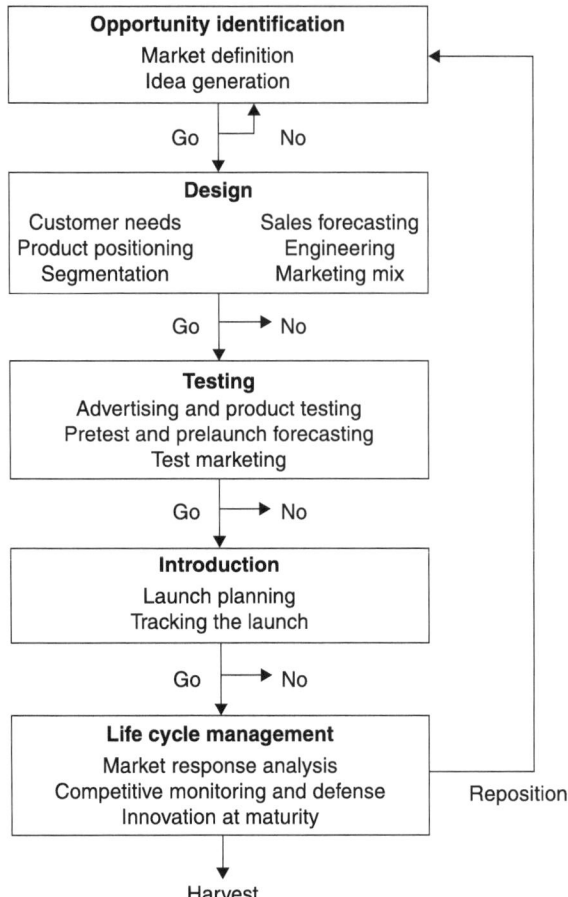

6.7 New product development process (Urban and Hauser, 1993: 38).

power. Urban and Hauser have also created a model for this complicated series of interactions, shown in Figure 6.8.

Even if all goes according to plan in the development of a new product, the consumer may still not accept it. Understanding the consumer adoption process will help a company know what to expect and enable them to plan accordingly. Consumers have been categorized according to their willingness to adapt a product. Understanding the diffusion curve for product adapters will help a company predict the success of their product. Figure 6.9 shows the diffusion curve, which is a visualization of the spread of innovation through the social system.

Fabric developments are highly dependent on the nature of the fiber or yarn component. Understanding the key requirements that must be fulfilled in order to

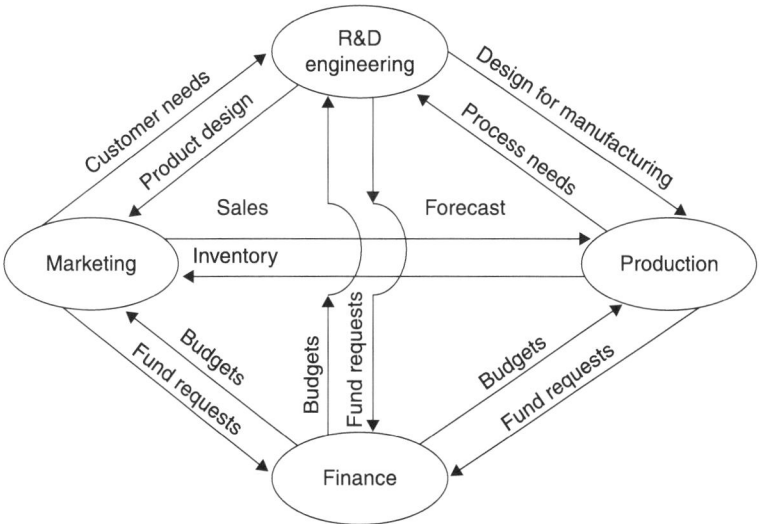

6.8 Cross-functional integration (Urban and Hauser, 1993: 33).

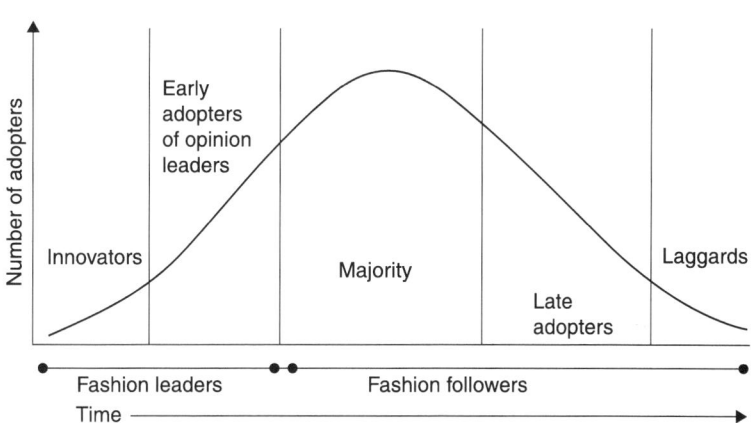

6.9 Product adaptation diffusion curve (Brannon, 2005: 43).

achieve a successful fiber-to-fabric engineering program is essential (Powell, 2004). Figure 6.10 outlines this process from a textile engineering perspective.

6.3.1 Factor-trend model

Eason (2009) developed a model of the relationship between decision-making power and the effect on automotive upholstery design trends. The fiber-to-fabric process was re-examined, with particular focus on the stages of design concepts, design analysis, design conceptualization, and material selection.

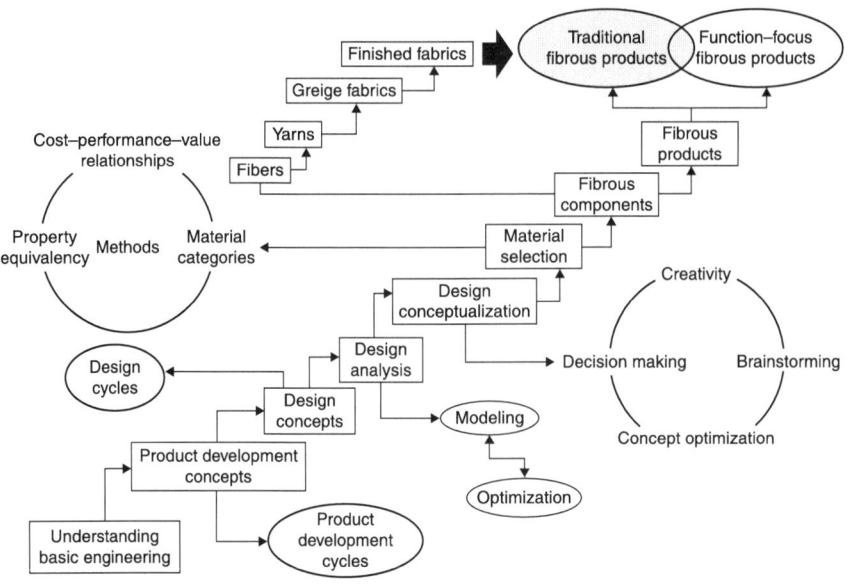

6.10 Fiber to fabric process (Mogahzy, 2009: 16).

The branches from each of these steps show cyclical relationships between brainstorming, conceptualizing and modeling, as well as selecting materials based on 'cost–performance–value relationships'. With consideration for these relationships, Eason's model reflects the product design trends resulting from the three primary groups of decision makers (design, engineering and purchasing). The model is a pie chart separated into three slices, which represent these groups of decision makers. Connected to each slice are descriptions of observed trends (specifically scale, motif and color), at the times that each role holds the decision-making power (Eason, 2009: 110). These relationships are outlined in the factor-trend model below (Fig. 6.11).

According to interviews with automotive textile industry professionals, these three branches of product developers have a powerful relationship. The best situation for OEMs, suppliers and the consumer is when all branches work in harmony. When the system is in harmony, designers' visions are achieved while meeting engineering standards and specifications, and staying under the purchasing price point. However, this harmony is difficult to realize and as one branch grows stronger, certain fabric trends take place. According to Eason,

> when Design has the majority of the "power", automotive upholstery fabrics are visually and technologically more innovative and achieve more variety in scale, motif, and color. A higher percentage of fabric motifs are large and medium scale organics and geometrics. Colors are generally high and low value with mid-saturation. Designs reflect increased inspiration from outside industries (home,

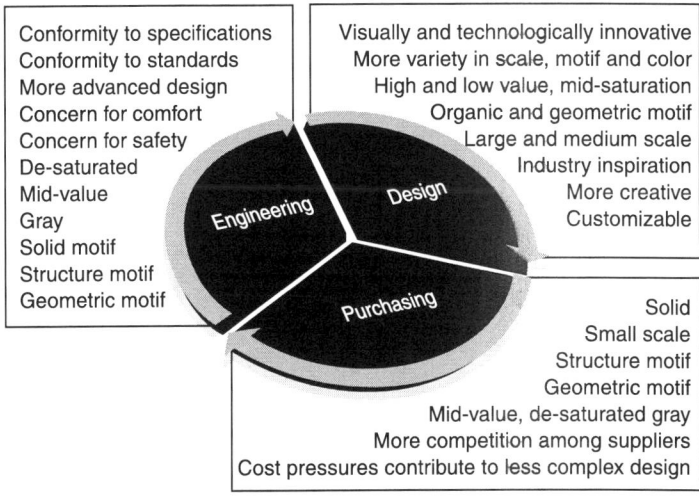

Conformity to specifications
Conformity to standards
More advanced design
Concern for comfort
Concern for safety
De-saturated
Mid-value
Gray
Solid motif
Structure motif
Geometric motif

Visually and technologically innovative
More variety in scale, motif and color
High and low value, mid-saturation
Organic and geometric motif
Large and medium scale
Industry inspiration
More creative
Customizable

Solid
Small scale
Structure motif
Geometric motif
Mid-value, de-saturated gray
More competition among suppliers
Cost pressures contribute to less complex design

6.11 Factor–trend model (Eason, 2009: 110).

fashion, electronics), are more creative and offer more options for customization. For example, Design held the decision-making power in the 1960s and 1970s . . .

When Engineering has more control over automotive upholstery fabric decisions, fabrics will, first and foremost, conform to strict standards and specifications and will focus on safety and comfort. Engineering is capable of achieving advanced design with new materials and technologies. Feedback from industry professionals reveals that fabrics are more likely to be solid or have simple, small geometric and structural motifs. A higher percentage of colors are gray, mid-value, and de-saturated. Engineering has played a significant role in automotive upholstery design decisions since the 1980s as government and industry standards for safety, comfort, and performance have continually increased, adding new pressures to Design along the way . . .

Since the 1980s, it is clear that Purchasing has held the majority of the decision-making power. This is reflected in the trends from this period to present day. When Purchasing is the "majority" power-holder, automotive upholstery fabrics tend to be solid or small geometric and structural motifs in mid-value, de-saturated grays. When Purchasing is in control, there is increased competition among suppliers to make fabrics for less cost, which contributes to less complex designs.

(Eason, 2009: 111)

The relationships presented in this model are applicable to a variety of technical textile industries and provide an opportunity for new product developers to predict trends once the decision-making relationship is understood.

6.3.2   Product life cycle models

Researchers across a number of different industries have attempted to understand product life cycles to create successful products. According to Mogahzy, 'product lifecycle analysis has become an essential task of product development in today's global market' (2009: 50). Figure 6.12 examines three types of product life cycles with very different rates of growth, saturation and decline. Brannon's life cycles visualize short-lived fads versus classics that stay in style for longer periods by graphing their popularity and duration (Brannon, 2005).

Laver attempted to explain the nature of fashion cycles with Figure 6.13. He proposed a 'gap in appreciation', the period during which old fashion looks are rehabilitated or consumers become comfortable with an innovative look (Brannon, 2005).

6.3.3   New product development effects on automotive upholstery

The dynamics of the supply chain and product development process contribute directly to automotive upholstery design trends. In the 1960s, seven or more

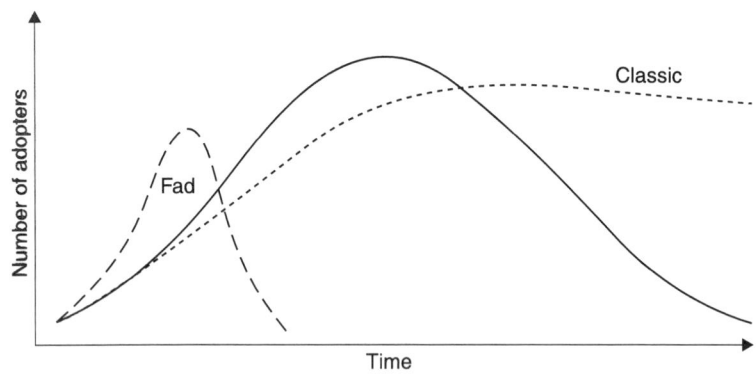

6.12 Fashion adoption trends (Brannon, 2005: 6).

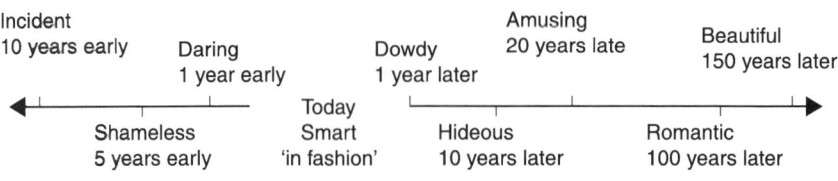

6.13 Laver's gap in appreciation (Brannon, 2005: 108).

automotive fabric suppliers existed in the US alone, and competed with each other through blind presentations of their fabrics to the OEMs. Originally, Tier Ones (seat makers) did not exist and fabric suppliers dealt with the OEMs directly. During this time the industry was much less global, and upholstery fabrics reflected localized design and taste. The industry relied less on data and research, and designers had the authority to influence the final fabric selection decision. Managers, engineers, sales and marketing did not necessarily consider trend cycles when deciding on fabric success, as long as the customer was 'satisfied'.

The structure of the industry began to change in the 1980s, when Tier Ones were introduced to work specifically with all seating components. As Tier Ones progressed, they wanted to work directly with the fabric suppliers. Fabric suppliers now had to not only please the OEM, but the Tier One as well. Additionally, Tier Ones began to focus strongly on the seating technology (frames, electronics, heating, etc.). Adding cost here drove fabric prices even lower. Purchasing began to have more control and began limiting the OEMs to specific suppliers. A 'bidding' process began among suppliers (OEMs asking suppliers who could make the same fabric at the least expense), resulting in lower quality fabrics and a disturbance to creativity.

It was also at this time that suppliers were required to become more global. As OEMs were producing more vehicles at international facilities, suppliers were expected to supply globally. Automotive upholstery suppliers began opening sales, design and production facilities in Japan, Germany, France, China, and the list continues. Collaborative design and development between design studios from the US with Asia and Europe contributed to a better understanding of the foreign OEM.

As the industry grew, so did the number of participants involved in the fabric selection process at the OEMs (demonstrated in Urban and Hauser's NPD Process (Fig 6.7)). A structured hierarchy developed where several individuals have the power to veto a design at any point. This filtering of designs led to 'watered-down' colors and patterns compared to those in the 1960s and 1970s. The design teams continued to lose power when it came to making design decisions. They were required to focus more on working within parameters set by purchasing and engineering, with influence from marketing. Although designers at both OEMs and suppliers made dedicated efforts understanding design and fashion trends from a variety of industries and understanding their vehicle's consumer demographics, their capabilities for making innovative and relevant fabrics could be overshadowed by one personal opinion in upper management, most likely related to cost or risk concerns.

Another important factor is the time it takes to produce a vehicle and a fabric. OEMs and suppliers look into fabric for a specific vehicle four years before the vehicle's production. Three years would be spent in the sample phase, two years in color development, and one year for everything to be tested and approved. Because this is an intricate process, and because of the decision hierarchy fabrics must go through for approval, changes cannot be made quickly, and therefore cannot respond accurately to current industry trends. This explains why trends in automotive fabrics follow trends in other industries such as fashion and home furnishings. By the 2000s, some suppliers even began to skip the design department at OEMs and go straight to engineering, purchasing or marketing for approval.

The entrance of imported vehicles (referred to as the New Domestics) has a significant impact on automotive upholstery design trends due to the internal structure at foreign companies. Foreign automakers have usually begun their entrance into the US market by building satellite design studios in the States, many of which are in Southern California, one of the most fashion-forward regions in the US. In this way they are able to understand the US market and consumers 'up close and personal'. Many of the designers at satellite studios, however, feel like they are more of a commodity and often find their office at odds with the OEM in the home country. The hierarchical structure at many Asian companies has been described as so intense that the company president is likely to 'point and choose' a fabric of his liking, without regard to US design input.

In the late 1990s and into the 2000s, when OEMs were causing suppliers to have 'bidding wars', they also required that fabric mills supply multiple technologies on a global basis in order to remain on their 'approved supplier' lists. In the more recent years, innovative technologies at suppliers have become the key to more advanced design and successful placements.

In today's global environment, all companies are looking at the same data (with the help of the Internet) and target the same customer. Many global companies are trying to introduce the same vehicle to various markets (Europe, North America, Asia and Australia, for example). This causes automakers to 'play it safe' and try to select fabrics that are not offensive in any market. Designers, however, know that all markets are not the same and no single style will satisfy all. Design is beginning to change in the face of the global environment and many designers insist that as the industry restructures, interior design will begin to be brand-specific, as it once was.

6.4   Novel materials and processes in automotive upholstery

The drive for increasing performance characteristics can be attributed to what is referred to as the 'NASA effect', where 'significant innovations developed for the military or aerospace industries are re-adapted for civilian use' (Colchester, 2007: 19). One recent automotive upholstery example was the introduction of 'smart seat fabric technology', which uses phase-change material, originally developed for NASA, to 'absorb and store excess body heat to create a cooling effect, and release it as needed to provide additional warmth' (Power Electronics Technology, 2004).

This section reflects the impact that the key drivers for automotive upholstery (standards and specifications, materials and technologies, economy and sales, and consumer and global influences) have had on current trends, novel materials and processes. The impact of environmental advances and computer-aided design are also discussed.

6.4.1   Current trends in automotive upholstery

Standards and specifications

Quality, performance and safety standards for automotive bodycloth have changed drastically since the 1960s and have also had an important impact on upholstery design. When there were fewer standards, changes could be made more quickly, which meant more time could be spent in the design phase and fabrics could continue to develop to meet aesthetically the changing appearance of a vehicle. By the 1990s, however, it took six months for production standards to be approved as specifications increased in number and demand. Also, if any design changes were made that changed the character of the fabrics, all tests would need to be redone, which meant that major fabric construction decisions needed to be made very early during a vehicle's development.

The 'NASA effect' of the 1970s caused an exponential increase in standards and specifications. As new developments from NASA government projects filtered into the automotive industry, the costs of meeting government standards also increased at an exponential rate. By the end of the 1970s, the automotive industry had the toughest and most stringent government standards, inside and out. As safety standards increased and the focus of automotive textiles moved to seatbelts and airbags, aesthetic design began to suffer.

Engineering requirements began to dictate scale, color and pattern, and ever-increasing standards challenged suppliers to create more durable, colorfast, lightfast and crocking-resistant fabrics. In the most recent years, consumers and OEMs are pushing away from mid-value for lighter fabrics as soil repellency and 'easy care' characteristics are increasing. At the same time, darker values are also reappearing with improved colorfastness, lightfastness and crocking characteristics.

Materials and technologies

The materials and technology available have limited automotive bodycloth. However, innovations in new materials and technologies have also opened many doors. Today, technology for fabric construction is limitless. As new technologies continue to be introduced, they will assuredly affect future trends. Fabrics being produced in today's vehicles are seeing a return to the large organics of the 1960s, but now larger than ever before. This scale is aesthetically exciting and has production advantages that hide trimming issues, bow-bias and sew lines. The industry is also seeing an increase in layers and secondary design processes (such as water-jet printing, laser etching and embossing) as ways to create interest on basic knits, flat-wovens and piles.

Through the late 1990s into the 2000s automotive interiors focused on technology beyond automotive upholstery, and suppliers began looking for new ways to add interest to fabrics without adding cost and complexity. Secondary processes on basic fabrics continued to increase, which contributed to trends in geometric, texture and structure motifs of all scales and colors. In the early 2000s OEMs were requesting more 'hi-tech' fabrics, which led to PVC-coated yarns, clear polypropylene yarns and fiber optics (non-production). Mesh and spacer fabric upholstery were also introduced in these years in small volumes.

Another technological advancement that has been in development since the late 1990s and has had a significant impact on value has been the introduction of new fibers, finishes and materials that are water and stain repellent, such as Milliken's Yes Essentials®, which allow for lighter interiors. Although this was first introduced for 'soccer mom' SUVs, which were sold in high volume into the early 2000s, all OEMs and consumers now expect upholstery with these capabilities.

Economy and sales

The 1960s were full of colorful geometric and organic motifs of all scales. This was a strong time for the US economy, and reflected a direct relationship of color and pattern with the state of the economy. This is due to the emotional relationship people have with color. In times of recession, designers push for more vibrant colors and patterns, which do not appear until three to four years later. Due to the nature of economic recessions in the US, by this time, 'happy' vehicles have been introduced to a strong economy. Approaching this issue from the opposite end, participants suggested that fabrics were plain, small and monotone during times of economic downturns. Regardless of whether design trends are 'happy' or 'depressed' all trends and all industry interviews, point toward one conclusion: cost pressures are 'king' and have a significant, overriding impact on fabric selection.

Although the economy was healthy in the early 2000s, the bright, happy colors and patterns of the early years could not return due to cost pressures. Instead, gray and beige solids and small geometrics continued to dominate the market. Due to the current economic crisis, and especially the travails of the automotive industry, US automotive upholstery fabrics have seen little change in the most recent years. Incremental changes are less risky than drastic changes. However, industry feedback also suggests that large geometrics and organics are on the rise, which will finally introduce this generation to something 'new and different'. In order to accomplish this, there is a push to standardize the bolster while changing the insert, which, as mentioned above, is a cost-effective way to add interest to vehicle interiors.

Consumer and global influences

Designers in almost any creative industry focus on making products that consumers can emotionally connect with in order to influence their decision to buy. This possibility has faded for automotive upholstery due to increasingly strict performance standards, internal decision-making struggles and cost pressures. Before these issues took their toll, cars were the epitome of American independence, self-expression and affluence. Vehicles became a way for people to express themselves and were offered in a variety of colors and styles with interiors to match. Automotive upholstery at this time was available in more colors and patterns than seen since, reflecting the 1960s and early 1970s as perhaps the only times that design held the power. As a result, vehicles reached out to their targeted demographics and were more fashionable. Automotive upholstery followed trends in apparel, home, music and art. Fabrics were bigger and brighter, like the bold primary colors of Pop Art.

Through the 1980s and 1990s, gray and beige dominated the market and began to reflect increased sophistication and 'smart consumer purchasing decisions'. The automotive industry was still marketing to the Baby Boomers, who held the majority of the nation's wealth, but as the population was maturing, so was design. Cars continued to reflect fashion, which at the time was corporate, sleek and gray. OEMs suggested that 'people want vanilla' and that if the interior were not understated, half of the potential consumers were going to reject or choose another vehicle (very few early adopters). It is important to remember, however, that consumers can only buy what automakers are producing. In addition, American consumers want instant gratification and buy at dealerships. As a result, they buy what is available. Dealers are hesitant to place any vehicles on the lot that a consumer may reject, and a one-off fashion color on the interior may be considered a risk to the dealer. Consumers were also taking more careful consideration of the financial obligations of owning such a big-ticket item, such as a car, which requires financing and as long as seven years to pay off the full cost. Cars became a necessity and a commodity. The emotional connection of the 1960s faded into utilitarian need.

The 2000s was a decade of technology and electronics, and interior designers were focusing on sound systems, GPS and many other exciting user-friendly technologies. Fabrics continued to be small, geometric, gray and simple. De-contented woven velvets moved to economy line vehicles and leather dominated the luxury market. Also contributing to more simple interiors, industry interviews suggest that the decreasing size of cars cannot handle bright colors and large motifs. Interest is added in the safest way possible, small and 'clean' texture and pattern.

Since about 2005, however, OEMs have begun focusing on brand differentiation and reaching a variety of demographics, rather than all lines at all OEMs focusing on a single customer. Generation Y is starting to have more influence, and as a result, more variety in scale, motif and color is expected to be seen in the most daring OEMs' new models. Black is also playing an increasingly important role, conveying sport and luxury, perhaps because it is the most expensive to dye and most difficult to meet performance standards. However, black is also the easiest color to recycle, which is important, as the focus on sustainability is becoming an influential consumer concern. As sustainability becomes more important to the consumer, more organic motifs and colors, such as greens and browns, which represent eco-friendly materials in most consumers' minds, are also beginning to emerge.

While just five years ago large patterns were considered too overpowering in small cars, designers now are focusing on small car fabrics with scales so large that, if there is a repeat, it cannot be detected. Vehicles and vehicle interiors are moving towards making more unique statements to better support vehicle brand character. Color is being experimented with in vehicles such as SMART and Scion, which are slowly answering Generation Y's requests for personalization. Industry feedback agrees on one thing: the US industry is currently making major changes that will affect the future of design. Most agree that until these tough economic times have passed and the automotive companies regain composure, little change will be seen in interiors. Upholstery fabrics will continue to be dominated by small gray and beige geometrics and plain solids.

6.4.2   Environmental advances

Since the 1990s, fabrics that in reality are more eco-friendly have been researched and produced in small volume, but not for production vehicles, due to either cost restraints or poor handling and durability. Now companies like Unifi are creating fibers and yarns that are made either from recycled content, are fully recyclable at the end of the product's life, or employ more eco-friendly treatment and chemical choices, and these are competitive with traditional materials in appearance, performance and handling, but still hold a higher price point.

6.5   Future developments in automotive upholstery

6.5.1   Factors predicted to affect future trends

Advancements in materials and technology are, perhaps, some of the most unpredictable factors. New technologies are developing constantly, and it is impossible to predict when and how the next will change the industry. However, in Eason's research (2009), industry professionals revealed technologies that are currently developed or in the development process. Many of these have been shown in concept cars for very high-end luxury vehicles, but have not yet become mainstream, primarily because of cost issues. They include pleats (or biscuits), foam embossing (or heat glazing), and various other secondary processes.

One of the most important factors to all of the above technologies, however, is seat design. Many automotive interiors continue to evolve with the goal of making decreasing vehicle weight. This is being done by removing foams, laminates and much of the seat structure, and replacing these materials with molded or suspended seats. The entrance of these technologies to US production vehicles is affected by the ability of the new materials to meet performance standards, price points and the acceptance of the car buyer. Changes to seat structure are predicted to affect design positively as seat features are expected to showcase design better and the interior is thought of as a whole rather than a series of parts.

Standards and specifications will continue to have similar impact on trends as before. Cleanability and recyclability will have increased contribution to future trends. As standards continue to rise for cleanability of automotive upholstery fabrics, the materials will be required to improve to meet these demands. These improvements will primarily affect the value of colors employed on interiors. It is predicted that much lighter-colored fabrics than ever seen before will begin to play an increasingly important role in automotive interiors, as will pure white. Many designers feel that as cars continue to get smaller, consumers will request lighter fabrics to perceptually create more open space. Light fabrics also convey luxury and quality to the consumer. However, until these materials are fully evolved and widely recognized, consumers will continue to be hesitant about purchasing a vehicle with a light interior for fear of dirt and stains.

As with past and current trends, future trends will be critically impacted by the economy, sales and the OEMs' willingness to take risks. Industry professionals predict that for the next few years, little will change in automotive upholstery design until the US economy settles and becomes strong again. Once the economy regains composure, fabrics are expected to be larger scale, both organic and geometric motifs, and available in a variety of colors. As an increasing percentage of fabrics become significantly lighter in value, another increasing percentage of fabrics will become more chromatic in the recovery from the current US recession. Also related to cost decisions, the percentage of solid fabrics is predicted to decrease as bolster fabrics become more standardized and are used in a variety of vehicles. This will allow for more focus on creating interesting insert fabrics to become the focal point of the seat.

As the global economy and automotive industry continue to recover, many changes will be made in the supply chain and among the power-holders. While this topic holds a high amount of uncertainty, its effect on future automotive upholstery fabrics will perhaps be the most influential. It is predicted that design, for the first time since the 1970s, may regain some control. If so, future automotive fabrics are expected to have more variety in all categories of design. Large organic and geometric motifs will reappear in a variety of colors, values and saturations. Each brand will begin to embrace its individual identity more strongly, as in the 1960s and 1970s. This will result in brand-specific colors and motifs.

In addition to customization to support brand identity, one of the biggest changes expected in automotive upholstery will be increased mass customization. Companies are beginning to focus more on their target consumer and are no longer targeting a single universal demographic. Influences of the youngest car-buying generation (Generation Y) are becoming increasingly powerful and their demands differ greatly from those of the Baby Boomers that automakers have been trying to please for the past 50 years. Generation Y has always lived in a world of computers, technology and global communication via the Internet. This generation is very design savvy and has a high demand for customization in order to express their individuality. Vehicles, such as Toyota's Scion brand, are now specifically targeting this generation, and are doing so through less expensive vehicles with more creative options. Generation Y is expected to influence automotive interiors that are bigger, bolder and more colorful.

Generation Y's desire for color and exciting patterns has been seen before with the young Baby Boomers in the 1960s and 1970s. Creating vehicles that met these desires proved very successful. Industry professionals reveal that much influence for future automotive fabrics will come from designs of the past, which, to the majority of the youth generation, will be 'new and innovative', and for the rest will be nostalgic of their parents' generation.

Another major influence on automotive upholstery design is an ever-increasing global awareness. The US automotive market has often followed the trends of the European market, which currently includes automotive fabrics that are big, bright and colorful. New colorways are being introduced through increased cultural awareness of Latin and South America. The US automotive market is predicted to become more colorful in order to satisfy the desires of the increasing Hispanic population. These factors, in combination with the entrance of China and India into the American automotive market, leave even the most experienced designers and trend forecasters unsure as to what is in store for the future of automotive interior design.

6.5.2   Forecasted trends

Because of the unpredictability of the factors mentioned above, industry professionals were much less consistent describing future trends than historical and present trends, as can be expected. According to Eason (2009), large-scale designs will dominate the US market by 2020, for the first time since the early 1970s. Mid-scale is also predicted to increase to about 15% by 2020. Industry professionals suggest that trends in color will experience dynamic change in the future. According to Eason, perhaps one of the most significant changes will be a downward trend of gray and beige hues. Color will begin to play a more significant role in automotive upholstery, but in new and innovative ways. As the variety of colors increase, industry professionals suggest that these hues will not be very chromatic. However, chroma trends also predict that as grayscale decreases and the least chromatic colors increase, more chromatic colors (bolder and brighter colors) are also increasing. In other words, future trends in color will be bi-polar. Trends in value tend to be more cyclical than any other set of trends investigated in Eason's study, which is predicted to continue into the future. Industry feedback suggests that black will continue to play an important role in automotive upholstery, as will very dark values. In 2015, however, the two lightest values will begin to rise to as high as 25% each, as the darkest value peaks and begins to decline. This increase in very light values coincides with increase in hue variety at low saturations.

6.5.3 Reducing the environmental impact

While polyester remains the dominant fiber choice in the automotive industry, increasing environmental awareness and recyclability standards have encouraged research into greener options for automotive interiors. Natural fibers, such as flax, Kenaf and wool, are being researched in combination with other man-made fibers, such as polypropylene (Powell, 2004). Also, according to Powell, because the traditional bodycloth laminate is 'composed of polyester fabric, polyurethane foam, and nylon scrim, it is difficult to dispose of or recycle. Attempts have been made in Europe, particularly in Germany, to replace foam with non-woven felt or spacer fabrics, but an appropriate substitute has not been adopted worldwide' (2004: 7).

As sustainable resources continue to evolve, the materials and technologies available will continue to have a big effect on automotive upholstery design. A large percentage of current research in automotive upholstery fabric and yarns is focused on creating more environmentally friendly products. As new yarns and fabrics evolve that are either biodegradable, made from recycled materials, or fully recyclable at the end of their life, their ability to accept various dyeing techniques will impact color possibility. While some industry professionals suggest that the sustainability movement will bring more perceivably natural colors (greens, browns, tans) and organic patterns, others suggest that this movement will lead to colors that in reality are more sustainable (such as natural fibers without dye or black fibers, which are easier to recycle, as previously mentioned). Industry professionals agree, however, that even by 2020 the cost for eco-products will still be too high to exist in large volume (although many suppliers are trying to change this).

6.5.4 Computer-aided design

Perhaps one of the most exciting advancements in automotive upholstery design and engineering is the ever-increasing prevalence of computer-aided design technology (CAD). Most OEMs and automotive upholstery suppliers are relying more heavily on CAD than ever before. According to Powell, 'Computer Aided

Design has been an effective tool in creating a range of designs and colorways more efficiently on paper and graphic simulations before the first fiber is dyed' (2004: 10).

6.6   Sources of further information and advice

Resources regarding the automotive industry are plentiful and exist through a variety of media, including academic and technical research books, journals, papers and periodicals, popular interests magazines, websites and videos, and government and corporate news and statistical data. *Ward's Auto, Edmunds.com, Interior Motives*, and *Car Design News* are just a few of the widely respected sources for automotive trends past, present and future.

Resources that focus on the textile industry are slightly less available, yet essential to developing an in-depth understanding of the factors that influence automotive textiles. *Engineering Textiles: Integrating the Design and Manufacture of Textile Products*, by Y. E. El Mogahzy (2009), is one of the most current and accurate depictions of the technical textile industry today.

Resources that focus specifically on the automotive textiles industry are few and far between. Perhaps this is due to the value of industry trade secrets. *Textiles in Automotive Engineering*, by Fung and Hardcastle (2001) is the earliest source used in this research and has been referred to as a 'bible' for automotive textile designers and engineers. Other valuable sources in this category include *Automotive Textiles*, by Adrian Wilson (2007), *Textile Advances in the Automotive Industry*, a collaboration edited by Shishoo (2008), and *OESA Industry Reviews*, prepared by the Original Equipment Suppliers Association (annually).

Additionally, Nancy B. Powell, Associate Professor at the North Carolina State University College of Textiles, has been a standout writer and specialist on the topics of automotive textiles, particularly upholstery design. Powell gained 20 years of industry experience with automotive textile suppliers prior to entering academia, and now shares her insider knowledge of her industry through numerous publications. Three of these, which focus specifically on automotive textiles, include 'Design driven: the development of new materials in automotive textiles' (Powell, 2004), *Transportation Interior Textiles: Function and Fashion* (Powell, 2005), and 'Design management for performance and style in automotive interior textiles' (Powell, 2006). This research is an expansion of these last three studies, which focus primarily on automotive upholstery design. Powell's collaborative research includes 'The development of woven velours for the transportation market', (Powell with S. Manley, 2004), and 'Automotive Textile Durability', (Powell with S. Rodgers and presented at AUTEX, 2006).

Powell's students have also contributed substantially to the automotive textile body of knowledge. Directly connected with this topic is Jenna M. Eason's Master's degree thesis, 'Factors affecting trend cycles in automotive upholstery design, 1960–2020' (2009), which reviews 50 years of past and present design trends, as well as a look into future automotive upholstery design, according to interviews with 70 industry professionals in design, engineering and purchasing.

For information regarding the NPD process this chapter primarily refers to the work of Urban and Hauser (1993), *Design and Marketing of New Products*, as well as Cagan and Vogel (2002), *Creating Breakthrough Products: Innovations from Product Planning to Program Approval.*

6.7   References

Anand, S. (2003). *Recent Advances in Knitting Technology and Knitted Structures for Technical Textiles Applications*. Istek.

Brannon, E. (2005). *Fashion Forecasting*. New York, NY: Fairchild Publications.

Cagan, J. and Vogel, C. M. (2002). *Creating Breakthrough Products: Innovations from product planning to program approval*. Upper Saddle River, NJ: Prentice Hall PTR.

Colchester, C. (1991). *The New Textiles: Trends and traditions*. London: Thames & Hudson.

Colchester, C. (2007). *Textiles Today: A global survey of trends and traditions*. London: Thames & Hudson.

Detroit Body Products, Inc. (1955–2006). *Detroit Book: Original Automotive Trim*. Wixom, MI.

Eason, J. M. (2009). Factors Affecting Trend Cycles in Automotive Upholstery Design, 1960–2020. Unpublished MSc thesis, North Carolina State University, Raleigh, NC.

Edmund's Automotive (2009). *Top 10 Best-Selling Vehicles, 1997, 1998, 2000–2008*. Retrieved from http://www.edumunds.com [Accessed 18 August 2011].

Fung, W. and Hardcastle, M. (2001). *Textiles in Automotive Engineering*. Cambridge: Woodhead Publishing in Textiles.

Grossman, R. and Wisenblit, J. (1999). 'What we know about consumers' color choices', *Journal of Marketing Practice: Applied Marketing Science*, 5(3): 78–88.

Johnson, J. (2005). 'Toyota sets lofty recycling goal for 2015', *Automotive News*, 6130 (January 17): 32F.

Manley, S. and Powell, N. (2004). 'The development of woven velours for the transportation market', *Journal of Textile and Apparel, Technology and Management*, 3(4).

Maynard, M. (2003). *The End of Detroit: How the Big Three lost their grip on the American market*. New York: Doubleday.

Mogahzy, Y. E. (2009). *Engineering Textiles: Integrating the design and manufacture of textile products*. Cambridge: Woodhead Publishing in Textiles.

OESA, (2008). *2007–2008 OESA Industry Review*. Troy, MI: McGraw-Hill.

Powell, N. (2004). 'Design driven: the development of new materials in automotive textiles', *Journal of Textile Apparel, Technology and Management*, 3(4): 4.

Powell, N. (2005). *Transportation Interior Textiles: Function and Fashion* (unpublished).

Powell, N. (2006). 'Design management for performance and style in automotive interior textiles', *Journal of the Textile Institute*, 1: 23–37.

Powell, N. and Rodgers, S. (2006). Automotive Textile Durability. Paper presented at the 2006 Association of Universities for Textiles World Conference (AUTEX), Raleigh, NC.

Power Electronics Technology, (2004). Ford, GM Try Smart Seat Fabric: Warmer in Winter, Cooler in Summer. Retrieved 25 September 2009 from http://powerelectronics.com/autoelectronics/ford_seat_fabric/

Shishoo, R. (2008). *Textile Advances in the Automotive Industry*. Cambridge: Woodhead Publishing in Textiles.

Urban, G. L. and Hauser, J. R. (1993). *Design and Marketing New Products* (2nd edn). Englewood Cliffs, NJ: Prentice Hall.

Ward's Automotive Group, (2009). *Ward's US Best Selling Light Vehicles, 1980–1996, 1999*. Ward's AutoInfoBank, Penton Media Inc.

White, L. (1971). *The Automotive Industry since 1945*. Cambridge, MA: Harvard University Press.

Wilson, A. (2007). *Automotive Textiles*. Silsden: Textile Media Services, Ltd.

7

Nanotechnology innovation for future development in the textile industry

F. NOOR-EVANS, KPMG – R&D Incentives, Australia,
S. PETERS, Queen Mary University of London, UK
and N. STINGELIN, Imperial College, London, UK

Abstract: This chapter discusses the process of 'de-maturity' of the European textile industry, moving away from its traditional roots in an attempt to revive the fortunes of this mature industry, through the adoption of novel technologies, such as nanotechnology, microelectronics and/or biotechnology. The process requires a paradigm shift involving every aspect of the firm that includes its technical capabilities, research and development (R&D) and business strategy. In particular, this chapter discusses a new product development strategy that permits the incorporation of the novel technologies into current business activities, which is consistent with the Open Innovation paradigm. This is illustrated through a case study of Freudenberg, a German textile firm.

Key words: textile industry, nanotechnology, innovation strategy, smart fabrics and intelligent textiles (SFIT), Open Innovation.

7.1   Introduction

The European textile industry has been in relative long-term decline since the end of the Second World War, precipitated by ever increasing levels of competition from countries with much lower labour costs. This is a classic competitive problem with a mature industry, which is characterised by standardised technology and requirements, and where cost is the main differentiating factor. Under these conditions, the competitiveness of the incumbent firms comes under immense pressure as the new competitors seize market share. One approach to halt, if not entirely prevent, such a loss of competitiveness is to implement a complete paradigm change and escape from the 'maturity trap' (Abernathy, *et al.*, 1983).

Indeed, the European Union (EU) has reinforced the importance of the need for a paradigm change with the introduction of the New Textile Technology Platform in 2004. The new platform identifies the incorporation of nanotechnology and micro (and flexible) electronics into textile products as one of the new growth areas for the European textile industry. The main objective of this chapter is to analyse the process of new textile product development through the adoption of the emerging technologies. This is a crucial issue as the technologies open up new opportunities to the textile industry given the fact that the industry as a whole, especially in Europe, is moving away from the traditional mainstream textile manufacturing in its attempt to find a sustainable competitive advantage. However, the adoption of the new technologies on its own is not sufficient. To thrive in this new environment, firms will find themselves obliged not only to develop new products and acquire new knowledge and skills, but also implement new organisational strategies and practices so that innovation becomes deeply embedded in their culture and routines. The latter is crucial because it will enable them to continuously change and adapt to an environment characterised by rapid technological and market change.

The chapter comprises six sections. The first section explains the various realised and potential applications of nano-embedded textiles as well as electronic textiles, and provides an overview of how those emerging technologies can create high value added products for the industry's consumers. The case study illustrates the successful adoption of nanotechnology and electronic textiles in an existing textile firm, Freudenberg, which incorporates the development of nanotechnology capabilities through the enhancement of existing capabilities, the role played by in-house research and development (R&D), and partnership with other firms. A number of challenges that may occur during the process of adoption are also addressed.

7.2   Nanotechnology in the textile industry

Nanotechnology is an emerging technology that has recently attracted a lot of attention as it promises to trigger a technological revolution that may affect industries across the board. The technology is expected to lead to unprecedented improvements in the performance of existing products, as well as the creation of entirely new products with specific properties that can only be obtained using materials with distinct features at the nanometre scale ($nm=10^{-9}$ m).

According to the Royal Academy of Engineering (2004), the term nanoscience is defined as 'the study of phenomena and manipulation of materials at atomic, molecular, and macromolecular scales, where the properties of materials differ markedly from those at a larger scale.' In a similar vein, nanotechnology 'represents the design, characterisation, production, and application of structures, devices, and systems, by controlling shape and size at the nanometre scale.' Thus, nanotechnology products are those derived from the phenomena and physico-chemical characteristics occurring at the dimension of less than 100nm (Siegel, 1999).

Such a definition leads to the view that nanotechnology facilitates the construction of products with building blocks that can be as small as a molecule, or a 'bottom-up approach' (Hu and Shaw, 1999). A case in point is bulk materials which can be produced with great precision, resulting in materials with the requisite properties, namely fewer defects and higher quality. Therefore, opportunities arise for modifying the fundamental properties of materials and designing structures with entirely new properties. In the case of textiles, nanotechnologies facilitate the improvement of existing textile properties, including ultra-hydrophobicity or hydrophilicity, high-performance flame retardance, chromatic behaviour and self-cleaning. In addition, the technology also enables the development of revolutionary textile products that include energy storage, the controlled delivery of drugs and Smart Fabrics and Intelligent Textiles (SFIT) (Qian and Hinestroza, 2004). The technology also enables complementary and conflicting functionalities to co-exist (for example, hydrophobic and hydrophilic), without damaging the original properties of the manipulated fibres or fabrics. In brief, nanoscience and nanotechnology open up a dazzling array of opportunities for the future development of the textile industry.

Various materials at the nano dimension have been introduced, such as nanofibres, carbon nanotubes, nanoparticles (oxide and metal nanoparticles), nanoclays and nanocapsules (Rao *et al.*, 2004, Wood and Scott, 2002, Qian and Hinestroza, 2004). A number of the micrometre-sized counterparts to these nano-systems are already commercially available and used in numerous commercial/consumer products. For example, synthetic microfibres have been used for filtration and industrial cleaning cloths by Freudenberg. International Flavor and Fragrance uses microcapsules for perfumery and Invista uses aloe vera encapsulated microcapsules in Lycra® filaments. In contrast, the majority of the corresponding nanomaterials are still at the R&D stage. Further investigation is required to fully understand the different characteristics and behaviours demonstrated by these materials before valuable commercial applications can be developed. It is equally important that efficient and effective manufacturing processes on a larger scale are designed, and that the health, safety and environmental issues that these entail should be addressed.

In order to produce nanotechnology textiles, nanomaterials need to be incorporated into textile artefacts. Several pathways exist that permit the integration of nanomaterials into textiles, including:

- *Incorporation of functional nanomaterials, such as nanoparticles, into fibres and/or on fabric surface.* This pathway is currently one of the most straightforward strategies to integrate nanomaterials into the existing textile manufacturing technology. For instance, functional nanoparticles can be pre-treated (dispersion in water) so that textiles can be impregnated or coated with the particles through a number of methods, including spray coating and the electrostatic method. For example, DuPont has introduced antimicrobial powders that are unaffected by solvents or a high temperature environment (up to 325°C), allowing them to be processed in a number of resin systems prior to spinning. An Australian firm, NanoShield™, has developed zinc-oxide nanoparticles in water that can be incorporated into fabrics or non-woven substrates through coating. The product derived from such a treatment is reported to impart UV blocking, anti-reddening and anti-inflammatory, antimicrobial or odour control properties to the treated fabrics. Another example is the self-cleaning fabrics developed by Schoeller through the modification of the fabric surface to mimic the surface structure of lotus leaves using silica nanoparticles.

- *Finishing technologies that permit the application of ultra-thin layers on fabrics or fibres.* Various promising approaches in this field are currently being investigated, one of which is plasma liquid (and gas) deposition. This method was originally a coating technology working in a low-pressure environment for micro-electronic etching. Following its introduction in the 1960s, within two decades its use has been widened to include different surface treatments, particularly in the field of metals and polymers. More recently, the textile industry has been seeking to adopt the technology to treat textile surfaces. For this application, however, the technology had to be adapted to allow deposition within an atmospheric-pressure environment. Plasmatex was the early European project (1997–2000) to develop plasma technology for flexible substrates operated in atmospheric pressures. Further progress on this front has been made by a number of textile firms in Europe, including Freudenberg that is planning to commercialise the modified technology in the near future.

Another finishing technology has been developed by Toray Industries, the largest textile and chemical firm in Japan. The firm made an advance in a processing technology that permits the formation of a functional coating of 10–30nm thickness, covering each of the monofilaments that structure the fabrics (woven or knitted fabrics). The approach, called 'NanoMatrix', is claimed to produce a uniform coating, thereby promising enhanced quality, performance and durability of the functionalities, a major improvement in performance, which the current coating technology lacks.

Japanese textile firm Teijin has developed a method to create a multi-layer fibre surface finish to obtain colour effects derived from light-reflection rather than the use of dyes or pigments. The multi-layer finish forms alternate laminates of polyester and polyamide of different refractive indices. As a result, the treated surfaces can produce coloration effects of an excellent visual quality that resemble the Morpho butterfly's wings when exposed to a source of light. Colour changes can be obtained through shifting the angle of the surface and changing the intensity of light. This innovation, called 'Morphotex', has been commercialised by Nissan for use in the interiors of their automobiles.

- *Encapsulating functional agents in polymeric nanocapsules prior to spinning or finishing processes.* Research in this field is focused on finding methodologies to control the release of agents, for example, for drugs, softeners and fragrances, in order to realise a continual and prolonged release. For the drug delivery application in particular, the controlled drug release can specifically target the intended areas, rather than randomly target them as is the case today (Textiles, 2005). For other applications such as dyes, sun-block agents and phase-change materials, on the other hand, research is being progressed to prevent the capsules from rupturing or zero release. It will considerably improve the quality and durability of the functions imparted by the treated fabrics. A number of commercial applications have been developed. However, these are based on the micro-sized capsules. One such application is Lycra® Body Care by Invista, which incorporates microcapsules containing different substances, such as Aloe Vera and Vitamin E, to Lycra filaments by means of padding or exhausting during wet finishing. Another application is the Ortho Evra skin patch that embeds birth control agents encapsulated in microcapsules (Textiles, 2005). Further improvements are now expected to materialise through the advancement of such technologies in nanocapsules because at the nano dimension the more capsules that can be incorporated into fibres, the greater the increase in performance.

In addition to the broad range of possible approaches to integrate nanomaterials and nanotechnology into textile production systems, various other technologies exist that allow for the creation of specific functions. One such alternative technology is derived from advanced printing technology which permits the application of functional materials onto the surface of fabrics or garments. Ten Cate, a Dutch technical textile firm, has recently acquired a British firm, Xennia Technology, which develops new inkjet technology to exploit inkjet digital printing technology for textile applications. The technology is being developed to produce functional fabrics, including solar cells, printed fabrics and protective fabrics, through means of printing. It allows for cost effective (accurate deployment of functional materials) and environmentally friendly (elimination of water) methods of deployment of nanosized functional materials on fabrics, something which cannot be done using traditional methods. Furthermore, the application of this technology in conjunction with innovation in dyes and pigments has potential for the development of smart fabrics. For example, fabrics that can change colour in response to different environmental conditions can be obtained through printing with chameleon-like dyes and inks. Other instances of responsive dyes and inks are thermochromics (sensitive to changes in temperature), photochromic (responsive to illumination) and solvatechromic (responsive to liquid) (Hibbert, 2004).

Smart textiles is another recent concept of advanced fabrics and clothing aimed at introducing functionality into textiles so that they can act as sensors and actuators with a particular stimulus operated by electronic systems. The textiles, classified as non-conventional textiles, are grouped within the SFIT category. This new category of textile is made of materials that can control their own functions according to changes in the environment, or that provide additional functionality, such as communication and physio-monitoring devices, with the use of electronic systems. This can be achieved through the treatment of the fabric core (the manipulation of fibres and yarns) or the integration of external components, for example, electrical current/switching, light energy through optics/panels, and sensors, such that the textile artefact can have the capability to sense its environment. The objective of research in the field is to add the desired functions into textiles and refine the performance without sacrificing other textile attributes with respect to the aesthetic, ease-of-care, durability and cost factors.

The potential applications of SFIT are immense, ranging from fashion, sports and home furnishing to military, outdoor, industrial, healthcare, medical and communication devices. Although the concept is mainly still at the R&D stage, a number of firms have commercialised early versions of wearable electronic products (Philips, Interactive Wear, Peratech, Eleksen, Voltaic System, Burton, Zegna i-Jacket and Adidas) for small niche markets. The current versions comprise two separate elements, the detachable wearable devices and the apparel itself. This structure allows the apparel to be washed without damaging the electronic circuits. An emerging approach is to generate circuit patterns by means of embroidery, weaving or knitting, thereby enabling the seamless infusion of electronics into the textile structure. This approach has encouraged research into conductive fibres for use as interconnectors and sensors. The current available conductive yarns (insulated copper wires woven into fabrics) lack the necessary performance including flexibility, reliability, durability and comfort features. Nanotechnology has the potential to fill this vacuum and address the weaknesses of the existing technologies. Carbon nanofibres and carbon nanotubes have shown remarkable thermal and electrical conductivity, in addition to being the most rigid and strongest fibre ever discovered. They are durable and flexible, and can be woven into fabric (Dalton *et al.*, 2003). Research in bio-sensing, using carbon nanotube-coated cotton threads to produce conductive cotton threads, is being conducted by a team at the University of Michigan. The threads can be woven into fabrics and are lightweight, non-corrosive and wearable. These are highly sensitive and can monitor the performance of a human's vital organs (Bourzac, 2008). The research, however, is still at a preliminary stage. The next stage of research will include the scaling up of the sample materials, integrating the conductive fibres into a communications device, searching for the appropriate applications, and measuring the cost-performance attributes of the product. The latter is a particular concern as carbon nanotube is significantly more expensive than other traditional conductive materials, such as copper. Therefore, scientists are looking into the advancement of conductive polymers to be spun into yarns.

This shows that SFIT involves an integrated system of complex elements with textiles as the platform. The concept requires an integration of knowledge and know-how that includes advanced materials, microelectronics and textile technology. Such a multi-disciplinary R&D requires a new approach in research collaboration, from which new knowledge and know-how can be generated. The new knowledge may act as a disruptive force to textile technology and the traditional functions of textiles and clothing; while the new approach creates a new cluster in textile R&D, which, when it reaches a certain critical mass, may change the structure of the industry. Therefore, this novel concept may foster the 'de-maturity' of the textile industry in Europe.

Indeed, recent R&D programmes in the European textile field has focused on encouraging a number of different players from various sectors to collaborate with textile firms to form networks of alliances. Opportunities that may arise from this emerging pattern of R&D have spurred new forms of collaboration and linkages between textile firms and other firms from different industries. From the sectoral system of innovation perspective (Malerba, 2002), this new development may cause a rapid shift in the industry's boundaries, demanding new firm strategies and business practices. Furthermore, textile innovation, especially in novel properties of textile functions and in non-conventional textiles, has the potential to change the entire landscape of the textile industry that includes the nature of competition, the supply chain and industrial structure. The new concept will require different levels of complexity in knowledge integration, technology and product development, and manufacturing and distribution systems from those required by the current structure. The new paradigm may also witness non-textile firms, such as electronics and polymer producers, invading mainstream textile markets. Smart textiles and intelligent clothing is therefore viewed as the next generation of textiles and clothing that may revolutionise the industry.

Although the potential economic and societal benefit of nanoscience and nanotechnology appears to be very promising, particular concerns have been raised about the implications of nanomaterials on human health, safety and the environment. For instance, silver nanoparticles are believed to possess powerful antibacterial and antimicrobial properties that offer benefits to human health. However, some particles may be able to penetrate the human body that could cause harm or may kill other microbes that are necessary for the environment. Due to the limited knowledge currently available about the adverse effects of nanomaterials and nanotechnology, precautionary measures need to be taken and research into the larger social impact of this new technology needs to be accelerated to match the pace of that in science and technology (Mnyusiwalla *et al.*, 2003; Royal Society). This movement is hardly surprising as a number of other new technological developments have come under close public scrutiny, an obvious example being genetically modified plants. Therefore, Europe and other developed countries have launched research projects to rigorously address the ethical, legal and societal issues of nanomaterials.

7.3   Adoption of nanotechnology for textile applications

The previous section argues that emerging technologies, such as nanotechnology and electronic textiles, have the potential to take the textile industry in a completely new direction, far beyond its traditional roots. Such a transformation will require a firm to reconsider how it is structured and behaves if it is to ride the 'revolutionary wave'. The following case study suggests that the adoption of emerging technologies will be fruitful only if a firm possesses the necessary levels of innovative capability and creativity to translate the technology into the innovative products demanded by customers. To achieve such capabilities and creativity the firm needs to possess a solid and comprehensive knowledge base and technology portfolio, as well as a deeply embedded culture of R&D and innovation. These factors will improve the absorptive capacity of the firm, helping it to appreciate the new opportunities brought in by the emerging technologies, which in turn encourage investment (Cohen and Levinthal, 1990).

The case study also shows that the adoption of a radical technology does not imply the need to discard current competences or that they will necessarily be destroyed. Interestingly, the newly adopted technology can be incorporated into the firm's existing knowledge base and technology portfolio and be used to create new products or considerably enhance the value of existing products. This is not to suggest that firms with specific competences cannot jump on to the emerging technology bandwagon. Instead, they need to focus their attention on developing networks of customers and suppliers, engage in collaborative R&D, partnering and alliances in order to build new technological capabilities and to create new products that meet the requirements of customers. For both small and large textile firms, collaboration, partnering and alliances are an important element in the technology adoption and commercialisation strategy. Quite simply, it can accelerate the building up of a new technological capability, accelerate the product development process, and exploit the advantages open to first-movers in the market (Teresko, 2005).

The case study of a German textile firm, Freudenberg, illustrates the dynamics involved in the adoption of emerging technologies at the firm level. Given the new paradigm the technologies offer, particular attention is paid to the strategy of the acquisition of new knowledge, the product development process to exploit the knowledge, the impact of the new technologies to the current and new businesses, and the management of change.

7.3.1   Freudenberg

Freudenberg is a family-owned business and was founded by Carl Johann Freudenberg. In 1848 he started his business empire by establishing a leather tannery business in Weinheim, Germany. Freudenberg has experienced a remarkable transition and emerged into a highly diversified firm, a world away from its original line of business. Freudenberg is one of the oldest and largest producers of nonwoven fabrics for a variety of applications, in addition to being a major global producer of seals, vibration control technology components and specialised chemicals (lubricants). Today, the company comprises 434 companies (made up of a variety of joint ventures and subsidiaries), it operates in 52 countries and employs approximately 32 000 people. Its headquarters remain in Weinheim.

Freudenberg is an interesting case study on the grounds that it has shown a remarkable ability to constantly change throughout its 160-year history. From its humble roots as a leather tannery, based along traditional craft lines, the firm has emerged into a large, globally diversified firm, and has established a solid record of initiating and managing continuous change.

This case study focuses on the firm's recent product development activities in the field of nonwovens. In 2007 Freudenberg's nonwovens generated sales of €1.03 billion with around 5000 employees. This success stems from the firm's commitment to take advantage of the latest scientific and technological advances to maintain its long-term competitiveness through technological leadership. Such a commitment enables the firm to offer a highly competitive product line-up to its customers in interlinings, filtration, hygiene/medical products, technical nonwovens, tuft and Evolon® (bi-polymer microfibre nonwoven) businesses, as well as to potential new markets. Freudenberg clearly illustrates, though, that technological innovation on its own is not sufficient to maintain long-term competitiveness. Rather, it has to be complemented with organisational innovation that facilitates the continuous generation of ideas, that is, the transformation of new ideas into commercial successes, supported by flexible management and production systems (Teece, 2007). The organisation has to be entrepreneurial and promote a culture of innovation. Accordingly, the case study is divided into two sections: technological innovation and the management of innovation.

Technological innovation

In 2007 Freudenberg employed over 1900 people in R&D, of which 1300 were located in Germany. In the same year, Freudenberg invested €202.9 million in R&D (including €2.6 million of government funding), or 3.79% of annual turnover. The figure is above the average R&D intensity in Germany which is approximately 2%. Nonwovens is one of the business areas that receive a significant portion of this funding.

In addition to internal research programmes, Freudenberg has been involved in collaborative research programmes, both funded by the German, US and Japanese governments (depending on the location in which research is undertaken by Freudenberg's subsidiaries) as well as by the European Union 6th Framework Programme for Research and Technological Development (2002–2006). Over the past six years, Freudenberg has received €32.3 million in government grants. Key research that involves textiles includes flexible electronic circuit boards, stretchable conductive substrates, fuel cell component technology and clean textile processing technology. Furthermore, collaborative research with universities, research institutes and companies has been a feature of Freudenberg's strategy from its earliest attempts at diversification in the early 1930s. In fact, this very strategy has prevented the firm from succumbing to the phenomenon of the 'maturity trap' as collaboration has improved the firm's technological abilities, whereby it continuously adopts new technologies and translates them into products in demand. In addition, Freudenberg is also a member of the Nanotechnology Competence Centre, organised by the Fraunhofer Institute for Material and Beam Technology. The centre facilitates the exchange of knowledge and synergies among its members, which include universities, research institutions, government and the private sector.

Freudenberg clearly illustrates that for the firm's long-term survival and competitiveness, alliances and collaborations are critical, amongst other things, to manage the enormous uncertainty associated with different technologies (Rosenberg, 1996). Powerful networks generated from alliances and collaborations may facilitate them to influence business ecosystems or to acquire some control over its environment such that the future of emerging market and technology, to some extent, could be defined. Furthermore, in an era of rapid technological and market changes as is the case today, hedging one's bet on a specific technology will certainly increase the likelihood of failure. Therefore, developing different networks to allow accesses to knowledge generated by and belonging to other firms is an attractive and viable strategy to manage the inherent risks associated with an emerging technology and market (Helfat et al., 2007). However, faced with limited resources, firms have to focus on a selected number of strategically important areas that will contribute to the firm's future competitiveness. One of the tools to help firms in this matter is technology and product road-mapping. The roadmap allows the firm to understand the future needs in particular markets and provides mechanisms to forecast, plan and coordinate technology developments to fulfil the needs. The roadmap provides general guidance/direction and shows opportunities for the firms to seize in the future. The role of entrepreneurial leaders is critical in selecting the strategic direction of research (Teece, 2007).

Armed with such an R&D strategy, complemented with a vast knowledge base and technology portfolio and long tradition in innovation, this is one of the firm's hard-to-imitate assets (Teece et al., 1997). The diversity of resources helps Freudenberg to develop cross-fertilisation and synergies between the firm's idiosyncratic assets and emerging technologies, from which unique new products and technologies derive. In the case of the adoption of nanotechnology, the firm's broad knowledge portfolio and experience provide the requisite absorptive capacity to develop various nanotechnology-related research projects that involve different nanomaterials (nanofibres, nanoparticles, and nanocoating) simultaneously (Cohen and Levinthal, 1990). Freudenberg has divided the application of nanotechnology in textiles into two pathways. The first was to maintain their leadership in nonwoven markets particularly in maturing segments, such as filtration, which led to production being relocated after 2005. The second was to utilise nanotechnology to create a new competence in surface treatment and perhaps in stretchable electronics.

Freudenberg envisages nanofibres as a complementary technology to advance their nonwoven high-tech filter technology for dust removal applications or liquid filtration up to membranes for bioreactors. The individual basic knowledge and technical know-how required to produce filters made of synthetic nanofibres (that is, nanofibre forming and finishing technologies), as well as filter design and manufacturing, has been accumulated within the firm for a number of years. The capability to form nanofibres has been developed in recent years as an advancement of their established capability in bicomponent splitable microfibre and electrospinning methods that have been acquired since the end of the 1980s. An opportunity to develop high value added products derived from the combination of those capabilities, arrived at with the emergence of demand for cleanroom filtration, was complemented by higher industrial standards introduced by the German government. The firm launched Viledon NanoPleat cassette filters for cleanroom filtration made of hybrid synthetic nanofibres (probably made of bicomponent fibres comprising polyester and thermoplastic materials in the form of sheath structure, core structure, an island structure or a side-by-side structure) at the end of 2007. The product sets new standards for indoor climate control technology, marrying fine filtration and high dust-holding capacity with energy-saving behaviour and longevity. Additional performance features, such as being corrosion-proof, highly resistant to chemicals, 100% moisture-resistant and microbiologically inert, were added to offer greater value to customers. Such a combination in performance, addressing primary and desired (secondary) requirements, is part of the product development strategy to offset the significant price increases associated with the introduction of new technology in products for existing markets.

Freudenberg appears to have foreseen the importance of functional surface treatment in the future, as well as the impact of the technology to some of its business areas, although the firm is not well-known for its competence in functional surface treatment. Therefore, in order to build surface treatment capabilities, Freudenberg teamed up with Dow-Corning – a firm that has considerable expertise in surface treatment technology – to co-develop plasma technology that can work in atmospheric pressure for coating flexible substrates. For this project, Freudenberg's advanced know-how in materials and component testing and in nonwovens processing technology, as well as its knowledge in polymers and chemical development, are complemented with Dow-Corning's competence in surface science. Several years of co-development has resulted in commercial atmospheric pressure plasma liquid deposition technology (APPLD), coating technology that enables modification of flexible substrates, such as nonwoven fabrics and polymer films, through extremely thin coatings of various materials that are chemically bonded to the underlying substrate. The technology is believed to facilitate novel, tailored surface functionalisation of a wide range properties, including waterproofing, low-friction slickness, adhesion promotion or antimicrobial. Plasma technology is claimed have the potential to produce very thin layers (within nanometre range), and to evenly distribute functional materials across surface areas without penetrating deep into fibres. As a result, functionalities can be significantly improved without adversely affecting the original fabric properties such as being breathable and lightweight. Furthermore, the technology complies with strict European environmental regulations as it is energy efficient, operates at near room temperature, uses no water, solvents or surfactants, and has negligible recycling needs. If the claim can be realised, the technology could potentially reduce the cost of production significantly, compared to the traditional wet-chemical methods, which consume a great deal of solvents, water and energy.

Diverse potential applications of electronics in textiles and stretchable products, which include healthcare, functional clothes and integrated electronics, are viewed by Freudenberg as the key domain crucial for its long-term growth. The firm possesses established knowledge in elastomeric materials, although its application in the electronics field was beyond the firm's competence. With an aim to acquire capabilities in stretchable electronics, the firm joins a collaborative R&D programme called STELLA (Stretchable Electronics for Large Area Application), a European research project under the 6th Framework Programme. Partners in the project include Philips (The Netherlands), CEA (France) and Technische Universität Berlin (Germany). The innovative products that may emerge from this project would be a result of the combination of knowledge in chemistry and materials with know-how in electronic interconnection and electronic system building. Several prototypes for specific applications have been developed, one being the 'sensing' of shoe inserts for patients with diabetes. Pressure sensors are being developed to give signals when the inserts need replacing (see Fig. 7.1). The system would assist in reducing

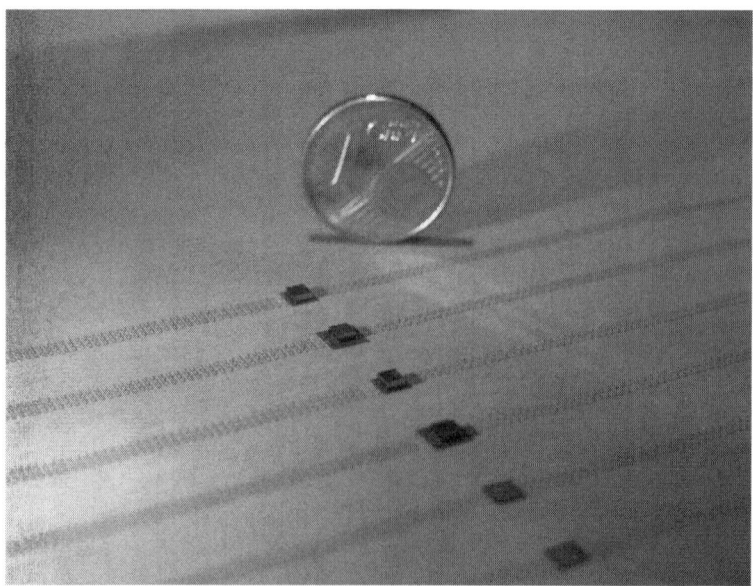

7.1 TPU test substrate with Cu conductors and dummy 8-pin component mounted from STELLA project. Source: www.stella-project.de.

the risk of inflammation of the feet and even amputation because of worn insoles. The electronics can be adapted for any shoe size and for each individual patient. For this application, Freudenberg sought a partnership with its former business group, Nora System, whose competence is in developing rubber floor coverings for a broad range of end uses, including homes for the elderly.

It is, however, far from clear which applications would lead the commercialisation of the new technology. However, having a number of partners within the consortium, each of which possesses different competences and capabilities, and are connected with various networks of customers and suppliers, viable potential end uses can come from various sources. This allows the network to target several emerging niche markets/applications, which could probably help to reduce the risk of overcapacity at the early stage of commercialisation or failure associated with an over reliance on one application.

It appears that the two different pathways for incorporating novel technologies for textile applications employ two different product development strategies. The first is to employ the novel technologies to significantly enhance the performance of existing products where the needs are known. Such a strategy is believed to be less risky as only one factor remains uncertain (the technological aspect), while the performance requirements are well understood. It allows the firm to build the necessary knowledge and capability whilst maintaining market risks at a lower level. Nevertheless, the performance enhancement should reach a certain threshold, at which potential customers would buy the new products, offsetting the likelihood of price increases. Therefore, for this strategy to be successful, collaboration with customers is vital. The lead user approach in developing commercial products pioneered by Von Hippel (1986) shows the significance of collaboration with 'lead users' in developing new products. Lead users are either small groups of users at the leading edge of important trends far ahead of a mainstream market, or 'trend-setters' in a particular market, who are able to articulate the likely future requirements of mainstream customers in the respective marketplace. This approach could reduce the risks of misunderstanding the most promising market, while building the capability in the associated technology.

The focus of the second strategy is on the creation of new markets. The complexity of managing new product development projects involving emerging technologies and markets seems to be greater. Where the behaviour of both technology and market is uncertain during the learning stage, the likelihood of failure in building new technological capability and/or in addressing the needs of new potential customers will increase. According to Wheelwright and Sasser (1989), failure in the process of developing new products often occurred in firms, due to misunderstanding the market, miscalculating their own technological strengths, or for that matter the technological challenges of the product. Therefore, partnerships, collaboration and alliances with customers and other suppliers are crucial to gain some control over the uncertain environment (Mattsson, 1988). Moreover, partnerships allow a range of different combinations of technology to be employed to address different combinations of product features. As illustrated by the STELLA project, this approach is also beneficial during market probing before the most promising market is defined. This strategy requires a team of people from different disciplines to work together as knowledge integrators and learning agents. Knowledge integration is a major part of new product development processes out of which difficulties inevitably arise. It involves a number of complex experiments and employs different problem solving techniques. Therefore, successful alliances obligate firm members to possess certain complementary assets and capabilities that allow interaction and transaction of knowledge among the members, from which unique, hard-to-imitate-capabilities can be generated (Helfat *et al.*, 2007). The management of alliances has to be able to balance contradictory elements of collaboration, amongst other things, between common objectives and individual firm's interests, and access knowledge that belongs to other firms. Thus, in addition to the contractual agreement, trust and reputation could be the safeguard of networked alliances (Helfat *et al.*, 2007).

The R&D strategy adopted by Freudenberg resembles the Open Innovation paradigm (Chesbrough, 2006). Chesbrough argues that the old model used by firms who conducted their own R&D, based on the premise it would guarantee their competitiveness, is now seriously out of date. The Closed Innovation paradigm, as it is otherwise known, has been shown to have failed firms badly in a number of high profile instances. In the competitive environment of the 21st century, characterised by an abundant knowledge landscape, an exponential rise in R&D costs and shorter technology lifecycles, the fundamental argument is that relying entirely on a firm's own research and competences is no longer adequate to provide a significant long-term competitive advantage on at least two counts. Knowledge has been increasingly distributed – as opposed to being concentrated within a few firms – which denies control over technological knowledge by a few large firms. Under these conditions innovative ideas come from multiple sources. Chesbrough argues these conditions were the main cause of the downfall of Xerox and IBM in the 1970s and 1980s. Secondly, in the era of multi-technological products and rapid technological and market change, the stakes are very high when investment is concentrated on just one technology, while investing in a variety of technological knowledge or platforms within an organisation is extremely expensive. Therefore, he suggests, firms need to work together with other firms (suppliers, customers), to acquire externally-generated technologies through either licensing or acquisitions.

The concept of open innovation suggests that firms need to share not only the cost of R&D and associated risks but also the outcomes arising from such undertakings. Such a changing R&D paradigm – from exclusive to distributed proprietary knowledge – certainly needs a shift in business strategy and the management of innovation activities. In developing business strategies, firms should take into account the criteria of suppliers, customers and competitors to be included in the networks, as well as the unique added value the firm would acquire from such networks that differentiate it from other competitors, particularly from those that are in the networks.

This new paradigm demands a certain type of capability, that is, dynamic capability (Teece, 2007), that facilitates the swift process of integration and assimilation of the knowledge into the firm's existing systems, and guarantees a continuous stream of new knowledge, ideas, and successful new products coming out from the process. Efficient and effective synergies between the internally and externally developed knowledge, and amongst the current knowledge and technology portfolio from which a firm can capitalise and maintain its long-term competitiveness, rely on the organisational capability of the firm. This issue is discussed in the following section.

Management of innovation

Freudenberg has learned from its 160-year history that long term success can be achieved through managing its innovative activities in such a manner that there is a continual stream of innovative new products coming out from the firm. However, technological innovation alone is insufficient to attain and retain long-term competitiveness. To make the product development process effective, firms need to build innovative cultures supported by entrepreneurial leadership, the necessary infrastructure (including networks) and resources and organisational capabilities. As Prahalad and Hamel (1990) argue, senior management needs to be firmly committed to innovation, demonstrate entrepreneurial leadership and devise the appropriate organisational structure, so that strategic assets are used efficiently to promote and sustain the competitiveness of the firm.

Freudenberg's corporate culture has been developed to encourage creativity and innovation, as it has been shown to be fundamental to ensure the future success of the firm. The culture was built from the bottom-up by empowering individuals, not only in R&D but also on the shop floor and the support areas. The firm provides the necessary facilities and infrastructure that allow individuals to leverage their knowledge and capabilities through training and development programmes, to communicate their innovative ideas throughout the entire organisation, and to join the teams working on new product development projects. In Freudenberg, innovative ideas do not necessarily come from senior management. In fact, a large number of new ideas are generated by the employees. Therefore, it is crucial that a simple and rapid process from idea generation to commercialisation is implemented. For that reason, the Management Board has introduced a series of innovation tools aimed at encouraging, accelerating, coordinating and measuring innovative activities.

This was implemented by Freudenberg with the launch of the Innovation Offensive programme in 2003 aimed at initiating specific measures for innovative activities, particularly in communication and methods. The result was the introduction of the Innovation Forum, a series of events created to facilitate communication and synergies among individuals and groups across business areas. It is a forum in which both individuals and groups can present innovative projects and invite other business areas to join the projects when required. The scope of the Innovation Offensive programme has been expanded continuously since 2004 to include strategy, R&D indicators and human resources. In 2004 the programme introduced an innovation monitor, that is, a method to assess its innovation portfolio and to estimate potential sales generated from innovative activities. In 2005 the R&D indicator – the number of patents, new products and process improvements – was introduced.

Following the introduction of innovation management procedures four years ago, the firm conducted a thorough assessment and there has been a subsequent realignment of its innovation strategy. The result of the assessment was the establishment of the Freudenberg Innovation Committee (FIC), whose members are the Chief Technology Officers (CTO) of all business groups, with an objective to identify key markets and trends in technology of interest to Freudenberg, and to foster the development of new products and technologies in the associated business groups. The FIC also leads the initiatives to benchmark the best practices of innovation management applied by other firms to monitor the efficiency and effectiveness of Freudenberg's own innovation processes. The progress is measured by the innovation indicators set up in 2005. In 2007 Freudenberg launched a new initiative to facilitate the development of an ideas pool and to accelerate new business development processes derived from the ideas. A bonus scheme, graded by milestones, was introduced in the same year, and a 'New Business Idea Award' is presented annually as a means of recognition.

In addition to providing infrastructure, facilities and tools for innovative activities, the firm develops routines to accelerate the progress from idea generation to new business development. It involves four milestones:

- Review by the Management Board, concerning whether the ideas are relevant to the interests of Freudenberg and whether the targeted market is ready to adopt the products derived from the ideas.
- Screening process by the FNT Board, involving two members of the Management Board and three Chief Technology Officers of major business groups. The basic criteria are:
  - potential sales (> €10 million after ten years)
  - whether the necessary basic knowledge exists within the firm
  - filtering fitness – where the ideas fit with one of the business groups, they ideas are then transferred to other relevant business groups. Otherwise, ideas are treated as projects.
- Scouting. Projects are given a limited budget to prepare documents that comprise business ideas, potential markets and sales, initial business plan, project planning and feasibility study, within three months. The results are presented to the Management Board.
- The project is extended to prototyping, customer acceptance tests and a thorough business plan.

For the complete process of new product commercialisation see Figure 7.2.

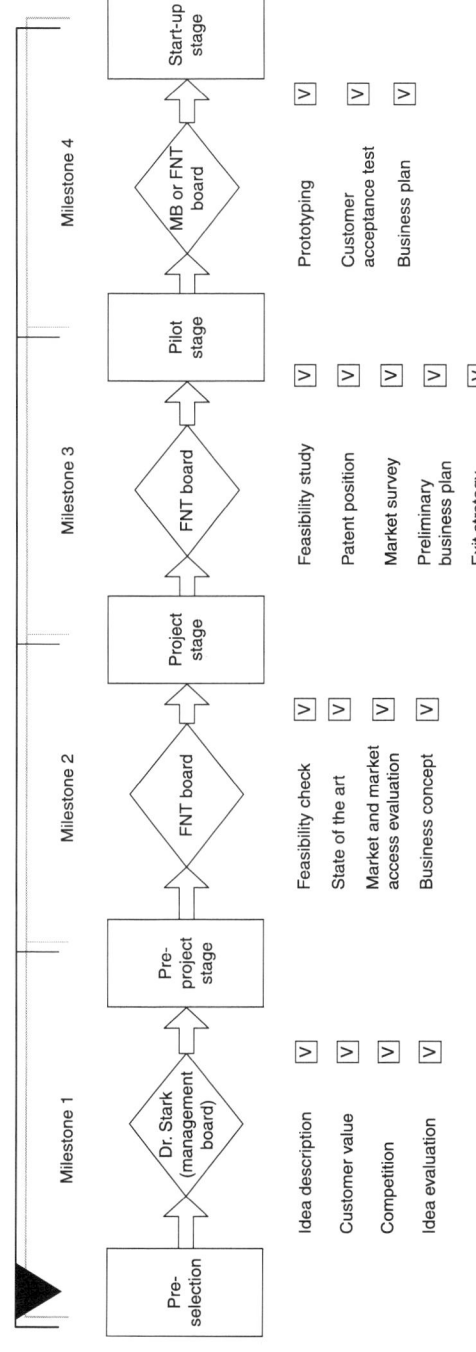

7.2 Stage-gate process in the business development unit. Source: Wendelken and Wolf (2008).

With an aim to harmonize activities related to new developments, Freudenberg set up a dedicated business unit to deal with business development for long-term growth. Since 2006 all previously autonomous innovation projects that were not related to any current business unit have been grouped under the Freudenberg New Technology (FNT) business unit. Its mission is to reinforce and sustain Freudenberg's strengths in innovation. The business unit focuses on consolidating group-wide technical know-how, especially that of interdisciplinary technologies, to develop new products and technologies. FNT are responsible for three main tasks: creating internal new businesses (internal corporate venturing); acquiring promising technology developers whose products fit Freudenberg's interests through Freudenberg Venture Capital (external corporate venturing); and partnering with relevant business units in developing new products and processes through the Freudenberg Research Service[1] (FRS, formerly Freudenberg Central Research). Other tasks include obtaining public funds, technology scouting and screening, internal and external network support, promoting the development of the next generation of technology professionals and creating an open and innovation-friendly atmosphere. Other business units focus on the evolution of their core businesses through product improvements and expanding the share of different markets, while the Management Board is responsible in providing the necessary guidance and consideration for acquisition. Currently, 95% of the R&D budget is allocated for improvements to existing businesses while the remaining 5% is dedicated to the creation of new businesses. Freudenberg's innovation activities are illustrated in Figure 7.3.

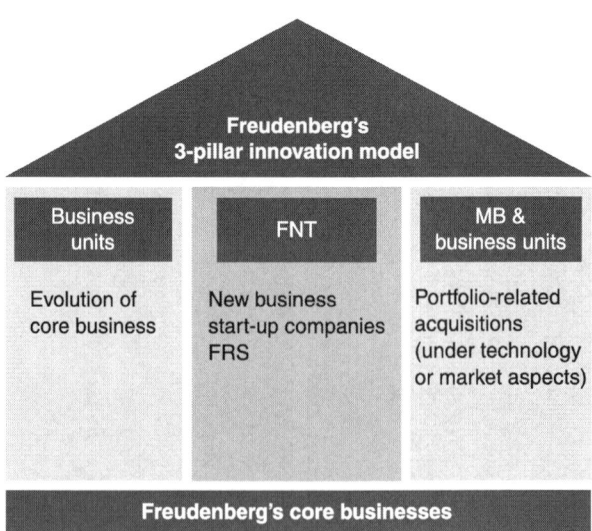

7.3 Freudenberg's 3-pillar innovation model. Source: Barth and Wolf (2007).

The business model adopted by Freudenberg shows that two different types of innovation – improvements and new creations/combinations (STELLA) or complementary technologies for different business groups (APPLD) – cannot be managed under one business unit. Individual business units focus on the evolution of their core businesses through improvements, while new creations or combinations are more likely to need a different set of knowledge and know-how, as well as different business models from those employed by the existing business units. Therefore, the latter needs to be managed by a dedicated unit, separated from the existing business unit. However, FNT and other business units have to have communication channels to enable synergies, from which new improvements and developments may be created. For that reason, the Management Board launched the Innovation Forum as discussed earlier.

To complement changes in business strategy and practices as a consequence of a shift in the R&D paradigm, firms also need to change their mindset when it comes to commercialisation. Firms can no longer restrict themselves to beating a single path to the market. Chesbrough (2006) contends that failure in commercialisation among high technology firms was, in greater detail, more affected by market uncertainty than by technical uncertainty. He suggested that Xerox's failure to commercialise a number of its cutting edge technologies can be attributed to the firm's inability to break away from its deeply entrenched knowledge of the traditional path to market and the need to serve its current markets. It lacked a systemic process for exploring and evaluating alternative business models. In Freudenberg, this is the role of the FNT.

7.4   Conclusion

A dazzling array of opportunities offered by the application of emerging technologies, such as nanotechnology and electronic textiles, are there for the textile industry to seize if it wants to in the future, although the ethical, legal and social implications of the technology need to be addressed. It is evident from the case study that there are a number of ways to adopt and commercialise emerging novel technologies in textile applications, either to improve the current products for existing markets or to create new products that are significantly different from traditional products. We argue that engaging with emerging technologies would enable mature textile firms to change the course of their strategic development. However, to help them to harness the opportunities, the Freudenberg case study shows that they need to change their traditional R&D paradigm, product development strategies as well as commercialisation paths away from traditional practices. Learning about the technology and the potential markets, either through internal development, acquisition, licensing, partnerships, alliances or the combination thereof, is essential as it allows the firm to leverage its knowledge and capability to build new technological capabilities. Therefore, the traditional practice, which assumes that a firm's competitiveness can be maintained based on a heavy reliance on internal R&D, appears to be outdated. In a similar fashion, the commercialisation of new products will need to exploit several different paths (start-ups, licensing, networks of alliances, partnerships with customers), depending on the business models that the firm believes to be optimal in the respected markets. It is, in fact, the open innovation paradigm which we argue will help firms to reduce the uncertainties associated with the emerging new technologies and markets. Therefore, in addition to technical capabilities, firms have to possess the organisational capabilities to create or acquire unique and difficult-to-replicate/imitate assets, manage changes in internal processes or routines, and select the possible paths the firm can take in the future. Support from senior management that promotes an entrepreneurial organisation is also critical. They are, in fact, the underlying factors of a long-term sustainable competitive advantage.

7.5   Future developments

The future development of the textile industry strongly depends on the capabilities of firms to break out of the maturity trap. One strategy is to adopt emerging technologies and develop unique products using a combination of established and new knowledge. For instance, the adoption of nanotechnology in the manufacture of textile artefacts promises one approach to create new opportunities away from the traditional textile roots. Significant government support, an increase in the R&D budget, and support from the private sector and research institutions in this field show an encouraging future for the European textile industry. A certain threshold nevertheless needs to be achieved to set in motion the industrial transition away from the traditional paradigm towards the new. In all probability this will affect every aspect of the industry.

According to Cientifica, given the potential performance enhancements and new products and markets that may be created, nanotechnology is expecting to provide incremental improvements to the traditional textile and clothing sector. The highest growth, however, will be generated from non-traditional sectors such as the military, sports textiles, medical textiles and aircraft, where performance rather than cost alone is the main driver. Nevertheless, the cost-performance factor will have to be taken into consideration as one of the basic requirements for the diffusion of nanotechnology in the textile market.

The market for nanotechnology-related textiles reached $13.6 billion in 2007 and is predicted to reach $115 billion by 2012. Meanwhile, Venture Development Corp predicts that electronic textiles, whose market in 2005 reached $380 million, will experience significant growth over the next few years. The growth will be generated from various markets including sensors (physiological status monitoring, cognitive status monitoring, movement/location), electronics (flexible and embedded processors/monitors/transmitters), textiles (conductive fibres, body armour, artificial muscles, biochemical hazard protection), energy (fuel microcells, photovoltaics, energy harvesting) and software/databases (embedded analytical software, databases with soldier physiological data).

In addition to facilitating improvements in performance and the development of new products, nanotechnology can contribute to the development of environmentally friendly products and production processes through reductions of materials, chemicals and waste. The main players in this market are Europe, the US and Japan, although a number of developing countries, such as China and South Korea, have begun to develop their nanotechnology capabilities. Europe is a clear leader in medical textiles, although in some other markets, such as military and sports, the region remains a major contender. A vast array of R&D is being undertaken. The challenge for firms will be to develop their nanotechnology capabilities in a timely fashion in concert with the technology and markets. Nevertheless, issues related to health, safety and the environmental implications of the technology still need to be taken into consideration.

7.6   Sources of further information and advice

The European Textile Technology Platform provides detailed information about the short, mid and long term research in the European textile industry. For discussion about the health, safety and environmental implications of nanotechnology, please refer to Annabelle Hett, *Nanotechnology: Small Matter, Many Unknowns*, Zurich, Switzerland: Swiss Reinsurance Company; The Royal Society and The Royal Academy of Engineering, *Nanoscience and Nanotechnologies: Opportunities and Uncertainties – Summary and Recommendations*, London: The Royal Society, 2004; and Wolfgang Luther (ed.), *Industrial Applications of Nanomaterials: Chances and Risks – Technological Analysis*, Dusseldorf, Germany: Futures Technologies Division of VDI Technologiezentrum GmbH, 2004. A number of books that examine the exploitation of novel functions in the textile industry have been published, such as Sabine Seymour, *Fashionable Technology – The Intersection of Design, Fashion, Science and Technology*, New York: SpringerWien, 2008 and P. Brown and K. Stevens, 'Nanofibres and Nanotechnology in Textiles', Cambridge: Woodhead Publishing, 2007.

7.7   Acknowledgement

We are indebted to Freudenberg, particularly to Mr J. Boecking and Mr A. Weghman, who have provided valuable information for the development of the case study. We would also like to thank the Goldsmith Company, Pasold Fund, University of London Central Research Fund, Prof Paul Smith (ETHZ) and Nanoforce Technology Ltd for their generous support. Any remaining errors are the Authors' responsibility.

7.8   Note

1   FRS is responsible to undertake research projects that have no direct links with the existing business units, or that require synergies across technological domains, including plasma coating technology and stretchable electronics. They are the research partner for all Freudenberg companies and subsidiaries in developing technology that has an impact across business units.

7.9   References

Abernathy, W. J., Clark, K. B. and Kantrow, A. M. (1983). *Industrial Renaissance – Producing a Competitive Future for America*. New York: Basic Books.

Barth, T. and Wolf, C. (2007). 'Developing the vision-tapping future markets', *FNT Info*, 1: 4.

Bourzac, K. (2008). 'Carbon-nanotube thread – fabrics woven from highly conductive, nanotube-coated cotton are wearable biosensors', *Technology Review Online*.

Chesbrough, H. (2006). *Open Innovation – The new imperative for creating and profiting from technology*. Boston, MA: Harvard Business School Press.

Cientifica (2006). *Nanotechnology for the Textile Market*. London: Cientifica.

Cohen, W. M. and Levinthal, D. A. (1990). 'Absorptive capacity: a new perspective on learning and innovation', *Administrative Science Quarterly* 35: 128–152.

Dalton, A. B., *et al.* (2003) 'Super-tough carbon-nanotube fibres', *Nature* 423: 703.

Helfat, C., Fingkelstein, S., Mitchell, W., Peteraf, M. A., Singh, H., Teece, D. J. and Winter, S. G. (2007). *Dynamic Capabilities: Understanding strategic change in organisations*. Oxford: Blackwell.

Hibbert, R. (2004). *Textile Innovation, Interactive, Contemporary and Traditional Materials*. London: Textile Innovation.

Hu, E. L. and Shaw, D. T. (1999). 'Synthesis and Assembly', in E. Hu, M. C. Rocco and R. W. Siegel (eds), *Nanostructure Science and Technology – A worldwide study*. Maryland: Loyola College.

Malerba, F. (2002). 'Sectoral systems of innovation and production', *Research Policy* 31(2): 247–264.

Mnyusiwalla, A., Daar, A. S, and Singer, P. A. (2003). 'Mind the gap: science and ethics in nanotechnology', *Nanotechnology* 14: 9–13.

Prahalad, C. K. and Hamel, G. (1990). 'The core competence of the corporation', *Harvard Business Review*, May-June: 79–91.

Qian, L. and Hinestroza, J. P. (2004). 'Application of nanotechnology for high performance textiles', *Journal of Textile and Apparel Technology and Management* 4(1): 1–7.

Rosenberg, N. (1996). 'Uncertainty and technological change', in R. Landau, T. Taylor and G. Wright (eds), *The Mosaic of Economic Growth*. San Francisco: Stanford University Press.

Royal Academy of Engineering (2004). *Nanoscience and Nanotechnologies: Opportunities and uncertainties*. Policy Document 19 R2.19, London.

Siegel, R. W. (1999). 'Introduction and overview', in E. Hu, M. C. Rocco and R.W. Siegel (eds), *Nanostructure Science and Technology – A worldwide study*. Maryland: Loyola College.

Teece, D. J. and Pisano, G. (1994). 'The dynamic capabilities of enterprises: an introduction', *Industrial and Corporate Change* 3(3): 537–556.

Teece, D. J. (2007). 'Explicating dynamic capabilities: the nature and microfoundations of (sustainable) enterprise performance', *Strategic Management Journal* 28: 1319–1350.

Teresko, J. (2005). 'From confusion to action', *Industry Week Online*. September 1.

Textiles (2005). Textiles and Drug Delivery. **3**.

von Hippel, E. (1986). 'Lead users: a source of novel product concepts', *Management Science* **32**(7): 791–805.

Wendelken, H. and Wolf, J. C. (2008). New business development – how Freudenberg wants to enter new business fields', *FNT Info* **1**: 1–8

Wood, A. and Scott, A. (2002). 'Nanomaterials', *Chemical Week*. Cover Story.

8

New product development in interior textiles

A. BÜSGEN, Niederrhein University of Applied Sciences, Germany

Abstract: New interior textile development is motivated by an ongoing global shift of supply and competition. Traditional manufacturers actually focus their development on the addition of innovative functions and utility value. This chapter describes in detail systematic procedural methods, which are used for product development of interior textiles. Four case studies illustrate successful product development examples for smart floors, acoustic damping tapestries, luminescent curtains and smart automotive cover fabrics. Learning experiences from these and other research projects are explained. The chapter closes with an outlook to future development trends and prospects of interior textiles.

Key words: interior textiles, innovation, smart textiles, idea rating, solution principal method, technology rating, cost effectiveness rating, smart floor, acoustic damping tapestry, luminescent curtain, smart automotive cover fabric.

8.1 Introduction

The term 'interior textiles' is often used as an alternative to the term 'home textiles', which sounds a little bit old fashioned. However, it does not mean the same. Interior textiles include textiles of vehicles like automotives, trains and airplanes, e.g. textiles for seating, ceiling or cover material of indoor panels. Architectural indoor textiles are also included into the term, as far as they are used indoor. Examples are acoustic damping textiles or screening fabrics for sun protection in any kind of building (Table 8.1).

Textiles, which are addressed by both interior textiles and home textiles are carpets, curtains, furnishing textiles, tablecloths, textile tapestries, blankets,

Table 8.1 Line-up of interior textiles

Interior textiles:
Curtains, window draperies and coverings
Wall fabrics (partitions), wallpapers (textile based), tapestries
Carpets, rugs
Beddings, bed sheets, mattress tickings, blankets
Upholsteries, cushions, furnishing accessories
Sunscreen fabrics
Towels
Tablecloths
E-shielding fabrics
Acoustic damping/absorption fabrics

132

bedding products and mattress ticking. Traditionally, the aesthetic factor is of major importance for interior textiles, because there is no other interior component that can accomplish a wider range of aesthetic features than fabrics and soft flooring (Nielson, 2007). However, the meaning of interior textiles changes and today, it is often more focussed on a technology-related viewpoint, which means that interior textiles include functionality (like fire resistancy, E-shilding, thermal compensation or light emission) and this aspect is frequently combined with decorative purposes and design. The Techtextil fair in Frankfurt, Germany, follows this trend and assigns technology-related interior textiles to a group called 'Hometech'.

It can be stated here, that the importance of performance and functional properties is constantly increasing for interior textiles. The examples described in this chapter will especially be of this kind of interior textiles.

8.2   New product development of interior textiles – basics and general procedures

There are some strong motivations for companies to develop new interior textile products. First of all, interior textiles have a great market share combined with an excellent growing perspective. The global market of home textiles was estimated in 2006 at US$105 billion, and a growth rate of 15% was expected from countries like India and China (Chakraborty, 2008). This positive market prospect arises mainly for the main export countries, which are China, Pakistan, India and Turkey.

In contrast, traditional manufacturers in Europe, the US and Japan are confronted with the on-going global shift of supply and they can only maintain competitiveness by new and innovative products. Their product development has to be different from standards and basic textiles. Those who decide to keep production in countries with high labour costs need more sophisticated products offering an additional value. There are two options for this:

- a highly ambitious aesthetic design, which may require expensive materials or an unusual and difficult to handle material combination;
- a novel function or an outstanding technological property of a textile which can be achieved either by new materials, by fabric construction or by modern finishing processes.

Many companies choose the second option and consequently have to undertake an expensive, time-consuming product development process full of risks and imponderability. Alongside a new competitive product, a new level of technology and knowhow is expected by companies who develop innovative products, and there are some practice examples where this expectation was fulfilled.

Another good reason to develop interior textiles for both new and traditional manufacturers is the fact that this kind of textile is not subjected to rapid changes, as are textiles in the apparel market, which have to follow seasonal fashion trends to be successful. The interior textile market offers more long lasting products, which may be improved over time, but not changed completely several times a year. Quality standards of long lasting textiles are much higher than those of short term use textiles and this leads to a high level of product value.

To sum up, the main motivations for new product development of interior textiles are:

- a huge market volume
- an interesting growing rate
- a global competition with low labour cost countries
- an addition of value (aesthetic or functional) to the product
- an increase of knowhow
- a long-lasting product characteristic leading to a high quality level.

The theory and concepts of product development have passed through a major enhancement, from the conversion of an idea, to a physical product, to a process, in which all product-related aspects, from raising an idea which fits to the basic strategy of a company, to the final phase of a product lifecycle, are carefully integrated. The use of powerful computer-aided tools has provided a new dimension in the process of product development, not only in the design phase of this process but also in other important areas including information gathering and marketing strategies.

The following list sums up the basic steps of a product development project, which could be an example of new interior or technical textile product development. This list does not exactly follow the appropriate product development theories published in economics literature (e.g. Guiltinan, 1997). It contains additionally well-founded practical experiences of textile product developments:

a. generation of a product idea
b. collection or working out of the relevant information
c. screening of the idea
d. conceptual design
e. manufacture of a product model, prototype or sample
f. testing and identification of the concept and product quality
g. development of the product production process and technology
h. marketing strategy
i. commercialisation of the product.

a. Generation of a product idea. In theory, idea generation should be systematic rather than haphazard (Tromsdorff, 1990). This demand, however, fails in daily practice. A great deal of idea generation is based on creativity. Systematic procedures may support a development project but can never replace creativity. The so-called 'flash of genius' is still an invaluable property of human beings. It may be inspired and supported by appropriate activities. Unfortunately, it cannot be forced to happen. The quality of human resources, as well as a positive atmosphere in a department or company, is responsible for the so-called 'internal source' of ideas.

Most important for the generation of new interior textile product ideas is to watch and to listen to customers. Customers frequently transfer their product idea to their suppliers to find out if this idea can be realized for reasonable costs. To benefit from customer product ideas, a high level of business confidence is necessary. After successful realisation of a customer product idea, some suppliers like to sell this product without restrictions on the market. In contrast, a customer wants to purchase this product exclusively. A solution to this conflict is to consider intellectual property rights and the quality of trust and relationship between the parties.

Analyzing competitor products is reported to generate about 30% of new product ideas (Trott, 2005). Especially for textiles, this is a well known but unpleasant procedure for getting new product ideas. Many interior textiles can only be protected by a registered design. Bypassing a design right is often too easy and, therefore, many companies have abandoned this option. Patents, which a company can more easily apply for with technically-focussed textiles, are a more powerful but expensive protection.

b. Collection or working out of the relevant information. The main purpose of this step is to prepare the following idea screening. Relevant information for a product idea can be cost analysis, forecast of product properties, availability of used materials, expected market acceptance, prices and production capabilities. Much interior textile development work takes place in this early stage to achieve precise and reliable information about the new product.

c. Screening of the idea. Screening is often combined with a rating procedure. As a result, the management should identify the chances of success, potential barriers and, last but not least, investment dimensions for the entire development process. Writing-up new product ideas enables a new product committee to review the idea. This write-up contains, for example, a product description, aimed market size, target customers, expected product manufacturing costs, achievable price, distribution channels, time and costs of the whole product development and rate of return.

The procedure of idea screening for a product idea is shown in Table 8.2, according to Kotler (Kotler, 1999). As an example for interior textiles, a woven car indoor panel cover having several smart functions (light emission for signals and illumination, textile based switches) has been selected. All relevant aspects of the product idea are listed in the first column. A weight factor represents the relative importance of the aspect in column two. Every aspect has to be rated now to determine how well the idea fits the capabilities and objectives of the company. Multiplication of ratings and weight factor leads to the results in column 14 (last on the right side), and summing up the aspect rating results in a value, which can theoretically be between 0 and 1. Some aspects may lead to a sudden finish of development without regard to other aspects, for example, if the market prospects of a new product are poor. For this reason, threshold levels of those aspects may be added into the table. This table shows three threshold marks (arrows) for

Table 8.2 Product idea rating of an indoor panel fabric having several technical functions (⇨ = threshold mark; ✗ = rating)

New product success factors:	(A) Relative importance	(B) Fit between product idea and company capabilities											Idea rating (A×B)
		0.0	0.1	0.2	0.3	0.4	0.5	0.6	0.7	0.8	0.9	1.0	
Company strategy and objectives	0.20	⇨									✗		0.180
Market chances	0.20						⇨	✗					0.120
Financial resources	0.15					⇨			✗				0.105
Distribution channels	0.15	⇨							✗				0.105
Production capabilities	0.1						⇨			✗			0.080
Research and development	0.1	⇨								✗			0.080
Suppliers' assistance	0.1	⇨									✗		0.090
Total	**1.0**												**0.760**

Source: Kotler, 1999.

market chances (minimum = 0.6), financial resources (minimum = 0.5) and production capabilities (minimum = 0.6).

d. Conceptual design. During product design, a large number of variations and alternatives arise and many decisions become necessary about which option should be discarded and which should be followed. The importance of reporting all relevant decisions and the reasoning behind them cannot be underestimated. Practice has proved that those decisions often have to be recalled and discussed and sometimes an alternative, which had not been selected to follow at an early stage, comes into focus later.

Searching for, selecting and combining the best options for a new product are a core activity of product development. Conceptual design starts with a list of tasks to be solved or questions to be answered for the new product (see the first column in Table 8.3). Next, a catalogue has to be created, which identifies the most interesting solution principles for each task. These solution principles are summarized in Table 8.3, where an indoor panel fabric development was split up into seven tasks with four principal solutions (A, B, C, D) for each task. A development team then had to find out and mark the best combinations of principal solutions.

Table 8.3 Systematic collection of relevant product design options for an indoor panel cover fabric having smart functions

Task:	rel. imp.	Solution principles to listed tasks			
			1	2	
Yarn material	0.1	Polyamide fibres	Polyester fibres	Glass fibres	Carbon fibres
Type of fabric	0.2	Non-woven	Woven	Warp-knitted	Weft knitted
Shaping of fabric	0.1	3D weaving	Cut and sew: manual tailoring	Cut and sew: Semi-automated	Thermal moulding process
Illumination technology	0.2	Light transmitting fibres	LED's	Electrolumi-niscent fibres	Fluorescent or phosphorescent fibres
Implementation of signal light	0.1	Conductive glue	Woven or knitted illum. threads	Soldering	Embroidering or sewing
Textile based switches	0.2	Integrated woven keys	Printed keys, capacitive sensor	Stitched keys, capacitive sensor	Double layer sensing fabric
Finish for easy surface cleaning	0.1	Polyure-thane-coating	Traditional hydrophobe finishing	Nano-finishing	Silicone coating
Sum:	**1**	**A**	**B**	**C**	**D**

This procedural method has its origin in the area of mechanical engineering construction and design of machines, where catalogues of solution principles (Roth, 2000) and a systematic evaluation (Pahl *et al.*, 2006) are common for many years. It has been transferred to technically-focussed textile product development by the author, taking into account the complexity of these products.

Table 8.3 lists four solution principles for each relevant task of a new indoor panel cover fabric. The decision about the right combination of principles still has to be done by the development staff. Two examples are given in Table 8.3. The first one (1) takes polyester yarn (option B), processes it into warp knitting (option C), processes it by semi-automated cutting and sewing (C), selects LED's as illumination technology (B), uses soldering for implementation of the required signal lights (C), realizes switches by printing and capacitive sensing (B), and gets a standard hydrophobe finish for easy cleaning properties (B). In a short form selection 1 can be identified by BCCBCBB. The other selection (2) is an alternative to this and can be identified by CBBBCDCC.

Evaluation and comparison of selections 1 and 2 are done by rating, which requires another list with valued importance and technology oriented ratings of principal solutions. This list is presented in Table 8.4.

Table 8.4 Technology rating calculation of solution principle no. 1 (see Table 8.3) for an indoor panel cover fabric having smart functions

Task	Relative importance	Rating calculation of solution principles (technology)			
			☑		
Yarn material	0.1	0.3×0.1=**0.03**	0.9×0.1=**0.09**	0.7×0.1=**0.07**☑	0.6×0.1=**0.06**
Type of fabric	0.2	0.3×0.2=**0.06**	0.9×0.2=**0.18**	0.7×0.2=**0.14**☑	0.6×0.2=**0.12**
Shaping of fabric	0.1	0.7×0.1=**0.07**	0.7×0.1=**0.07**☑	0.8×0.1=**0.08**	0.6×0.1=**0.06**
Implementation of signal light	0.2	0.4×0.2=**0.08**	0.9×0.2=**0.18**	0.7×0.2=**0.14**☑	0.5×0.2=**0.10**
Illumination technology	0.1	0.3×0.1=**0.03**	0.9×0.1=**0.09**☑	0.8×0.1=**0.08**	0.2×0.1=**0.02**
Textile based switches	0.2	0.6×0.2=**0.12**	0.8×0.2=**0.16**☑	0.9×0.2=**0.18**	0.5×0.2=**0.10**
Finish for easy surface cleaning	0.1	0.4×0.1=**0.04**	0.8×0.1=**0.08**	0.8×0.1=**0.08**	0.2×0.1=**0.02**
Solution	1 (sum)	**A**	**B**	**C**	**D**

The technology of selection 1 (selected principles BCCBCBB indicated with ☑ in Table 8.4) is calculated as:

$$0.09 + 0.14 + 0.08 + 0.18 + 0.08 + 0.16 + 0.08 = 0.81$$

The technology of selection 2 (CBBCDCC) is calculated as analogous to 0.74. Compared with the maximum possible rating of 1 and according to experience, it should be noticed that, basically, both selections may be followed to realize the new product. Selection 1 is rated higher than selection 2, but this rating is only taking the technology into account.

Cost effectiveness is another important aspect and has to be determined in addition for both selections. To this the expected costs of material and manufacturing have to be rated again for all solution principles. An example of this cost rating is given in Table 8.5. The cost effectiveness of selection 1 can be calculated now as 0.75, whereas the cost effectiveness of selection 2 is 0.72.

A relative importance of technology and cost effectiveness could now be defined in order to calculate an overall rating value for both selections. This is recommended, if too many selections are to be compared. In the given example, selection 1 is rated higher for both technology and costs and would be preferred for the next step of product development.

A systematic conceptual design should lead to a clear concept and a detailed version of the new product. It can be used additionally for strategic considerations, for eample, to compare the concepts of optimized technical and optimized economical solution principles with each other.

e. Manufacture of a product model, prototype or sample. This step is another core activity of product development, especially for interior textiles, and leads in most

Table 8.5 Cost effectiveness rating calculation of solution principles (see Table 8.3) for an indoor panel cover fabric having smart functions

Task	Relative importance	Rating calculation of solution principles (cost effectiveness)			
Yarn material	0.1	0.8×0.1=**0.08**	0.9×0.1=**0.09**	0.8×0.1=**0.08**	0.3×0.1=**0.03**
Type of fabric	0.2	0.9×0.2=**0.18**	0.7×0.2=**0.14**	0.8×0.2=**0.16**	0.6×0.2=**0.12**
Shaping of fabric	0.1	0.5×0.1=**0.05**	0.5×0.1=**0.05**	0.7×0.1=**0.07**	0.9×0.1=**0.09**
Implementation of signal light	0.2	0.7×0.2=**0.14**	0.5×0.2=**0.10**	0.8×0.2=**0.16**	0.9×0.2=**0.18**
Illumination technology	0.1	0.7×0.1=**0.07**	0.5×0.1=**0.05**	0.8×0.1=**0.08**	0.9×0.1=**0.09**
Textile based switches	0.2	0.7×0.2=**0.14**	0.9×0.2=**0.18**	0.7×0.2=**0.14**	0.3×0.2=**0.06**
Finish for easy surface cleaning	0.1	0.8×0.1=**0.08**	0.7×0.1=**0.07**	0.6×0.1=**0.06**	0.8×0.1=**0.08**
Solution	**1 (sum)**	**A**	**B**	**C**	**D**

cases to a sample or prototype, which may be tested, evaluated or presented to customers or to the public. Not only realization of the idea, but also optimization of product quality, manufacturing methods and the costing of materials and production are done at this stage.

For interior textiles, it is possible to use CAD to simulate the expected result in the form of a picture and to use this picture for marketing. Image processing textile CAD software offers realistic views of a textile product. Even the finishing of fabrics can be simulated today to show the visible effect at fabric surfaces (Hardt, 2006). The following concept testing may benefit from a simulation.

f. Testing and identification of the concept and product quality. Concept tests are usually done before prototyping and help to gain insight into marketing challenges and requirements when the product will be commercialized. If new interior textile products are developed, a large number of competing concepts may occur. Product features and visualizations are presented to potential buyers and structured questionnaires enable an analysis of customer reactions.

A textile is identified by touch and tactile surface properties. Therefore, only a hardware sample can be a fully satisfying approach to concept testing. Chemical and mechanical tests of product quality and safety help to create detailed technical descriptions of the new product. If important required properties are not fulfilled, the product concept has to be reworked.

g. Development of product production process and technology. If prototypes or samples are created by laboratory equipment, this step is of major importance and sometimes needs more time and financial resources than any other step of product development, because textiles are manufactured mostly as a large serial production.

Fortunately, samples of yarns, fabrics and sewn end products are frequently made with ordinary production machines, so it is not too difficult to adjust those machines to a serial production. In contrast, finishing and coating is often done more easily with laboratory equipment and, therefore, these processes require more work to be transferred.

h. Marketing strategy. The extent of test marketing usually depends on the costs of new product development. If these costs are high, a company likes to test the product and its entire marketing program in real market situations. That includes positioning strategy, advertising, distribution, pricing, branding and budgeting.

i. Commercialization of the product. After determination of the right time to introduce the new product, the manufacturing facilities have to be put into operation. Next, the location for the launch of the product has to be determined. It may be a single location, a region, a national market or the global market and this decision may be affected by the spendable budget.

Final remarks for the development of interior textiles. Out of any given 100 proposed new product ideas, on average, 39 enter the product development process, 17 survive this process, 8 make it to the market and only 1 is able to fulfil its business objectives (Tromsdorff, 1990).

Textile firms around the world are under constant pressure to become more efficient through technology upgrades. The driving factor of innovative interior textiles is the 4% annual growth rate of speciality fabrics. (Rasmussen, 2008).

8.3   Case studies

8.3.1   Smart Floor (SensFloor®)

The following example demonstrates that market success of a new and innovative product sometimes requires an extraordinary amount of work and time. The development of the smart floor called SensFloor has its origin in 2001, when Infineon, a Siemens spin-off founded in 1999, started to work on smart textiles. Integrating electronics into a carpet was the aim of a joint venture between Infinion and the carpet manufacturer Vorwerk. Awarded by Techtextil/Frankfurt in 2003, Infineon planned to present a marketable solution of a smart carpet by the end of 2004 (Infineon, 2003). At the end of 2005 commercializing was still not realized and Infineon closed down its activities in the area of smart textiles.

However, the leader of the development team, Mrs Christel Lauterbach, continued the development work and, in 2005, founded a company called Future-Shape GmbH. Another five-year period of development took place. During the development of Infineon, a team of six people, two PhD students included, had worked on smart textiles. From 2006 until today, seven people continue the work, assisted by additional trainees and students of local universities.

They had to optimize and change the smart carpet system and, in addition, develop a manufacturing technology resulting in the construction of machinery for it. The actual production machinery is based on fabric inspection units, having additional bonding equipment to integrate the electronic components into the textile structure.

A major change to the smart carpet was invented in 2006. The signal lines, which were textile based and integrated into the floor, were replaced by a radio transmission technology. Now, sensed signals are sent with a frequency of 868 MHz and a maximum operating distance of 30 meters to a receiving unit. As the application can be extended from carpets to many other floor cover materials, it is now called a smart floor system.

The SensFloor project has created a textile based underlay of floor coverings with integrated capacity proximity sensors and microelectronic radio modules (Fig. 8.1). The radio module is placed on a flexible printed circuit board and is designed to process data of up to eight sensor areas. Each square meter carries 32 sensors, connected by a delineated pattern printed onto the underlay. Activation of the capacity proximity sensors is caused by the walking or falling down of a person (Fig. 8.2). A sequence of location and time specific sensor events is broadcast to and received by one or more wireless control units (Federal Ministry of Education and Research, 2010).

SensFloor can operate mechanisms such as automatic doors, alarm devices, lights, heating or traffic counters. Automatic alarm triggering for people in hospitals, rehabilitation clinics or foster homes is one of the most attractive applications, because the staff can be alarmed for help, for example, when dementia sufferers try to leave or when bedridden patients try to get up by themselves.

A variation to the rolls of one meter width to cover the whole area of rooms or floors (Fig. 1) is a sensitive mat, which can be placed before an entrance or a bed (Figs 8.2 and 8.3).

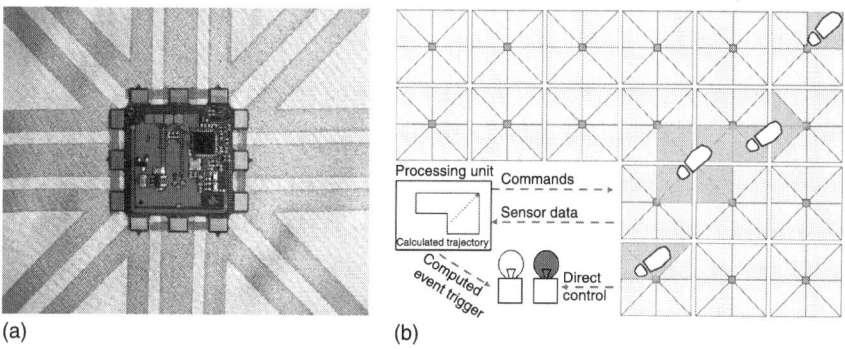

(a)   (b)

8.1 Future Shape radio module (a) placed on a textile underlay with printed signal lines and (b) triggered footsteps creating sensor events (Photo: Future-Shape GmbH).

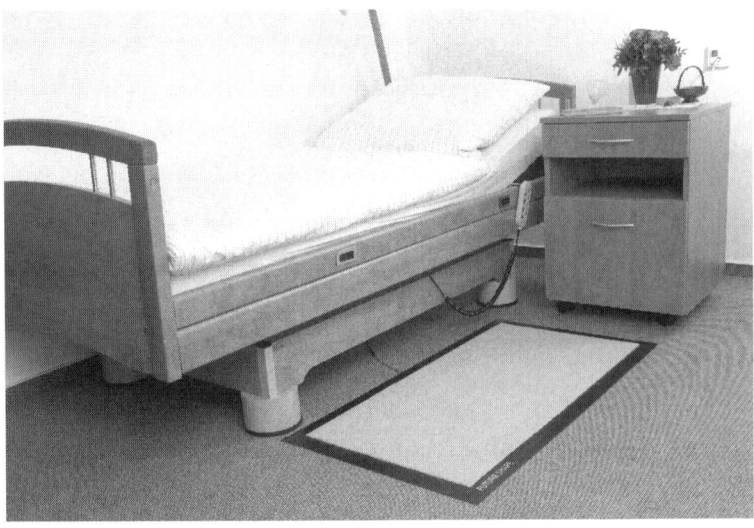

8.2 SensFloor mat uses 16 capacitive proximity sensors to detect the walking away or falling down of a person (Photo: Future-Shape GmbH).

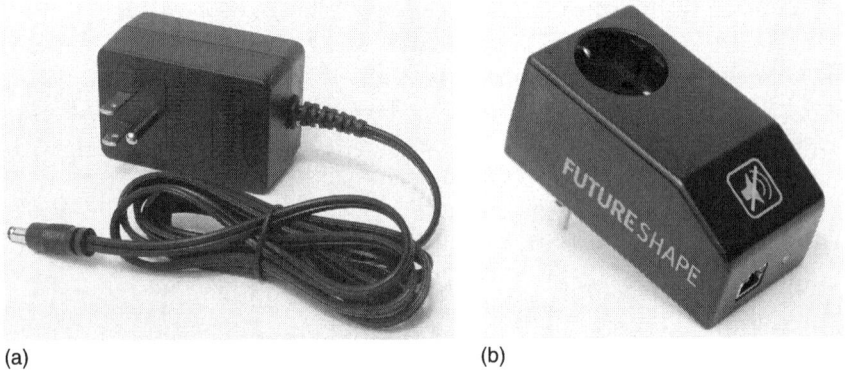

(a)   (b)

8.3 Power supply (a) and receiver (b) of SensFloor Mat (Photo: Future-Shape GmbH).

Commercialization of SensFloor started in May 2010. During the first 5 months, 12 systems have been sold to foster homes and 20 systems to private customers. It is a smart textile based contribution to Ambient Assisted Living (AAL). In future it may, of course, enter other markets as well.

8.3.2   Artistic tapestry with acoustic damping functions

The following example was initiated by an artist who looked simultaneously for a modern tapestry technology offering acoustic damping properties and the aged appearance of ancient tapestries.

When the new Cathedral of Our Lady of the Angels was being built in Los Angeles in 1999, the artist John Nava was asked to create a series of 57 large tapestries, each having a width of 2.3 meters and a length of 6.4 meters. He created a series called 'Communion of Saints'. After some weaving trials with unsatisfying results, a textile agent put John Nava in contact with the Belgian company Flanders Tapestries b.v.b.a., which is highly specialized at weaving tapestries and wall carpets. Nava found that this company could offer a heavy and high-class quality fabric.

The product development lasted about two years. Besides acoustic damping, very specific artistic requirements had to be fulfilled. The background walls of the church were to simulate Italian frescoes. The tapestries should not only match the background walls but also have an aged and antique appearance. Therefore, individual yarn dying was necessary and, to attain the correct colours in damped yellowish alabaster light, samples had to be measured in original light conditions with a colour spectrometer. The yarn colour was then adapted accordingly (Insider, 2001/2002). Additionally, cotton material was complemented by a minor percentage of viscose (Roman Catholic Archbishop of Los Angeles, 2008) and woven by a 3-layer weave.

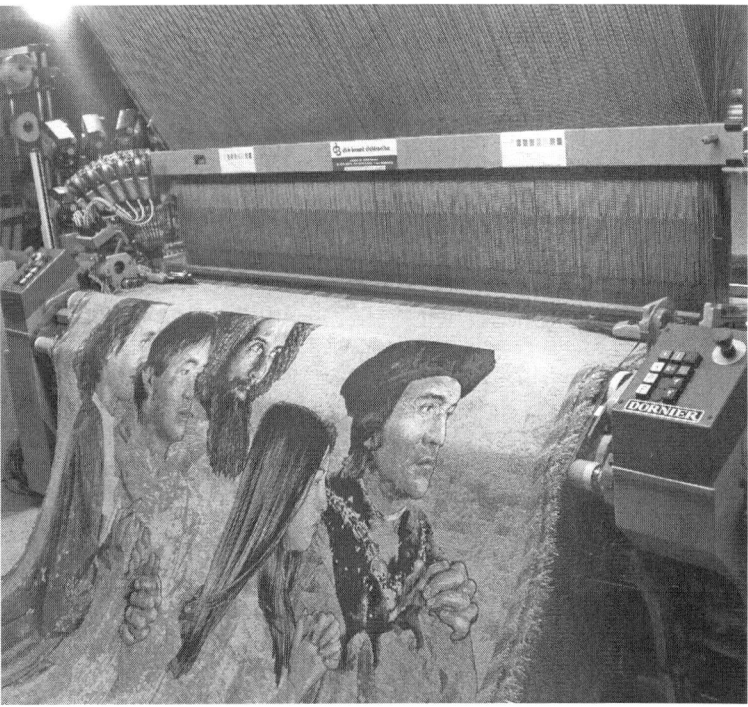

8.4 Weaving machine for tapestry fabrics (Photo: Lindauer Dornier GmbH).

(a)   (b)

(c)

8.5 Aged and antique appearance of woven tapestries created by sophisticated design, dying and weaving techniques (Photos: Lindauer Dornier GmbH).

Realization of the tapestries required one of the most complex weaving technologies in the world. A rigid rapier weaving machine using a gentle positive weft transfer (Dornier system) was equipped with a colossal Jacquard machine operating 20 000 lifting hooks to handle the 17 800 warp-ends of the tapestry set-up (Fig. 8.4). To transfer the design into a fabric, especially to solve tasks like clean-up and colour reduction, Flanders Tapestries had to develop specialized CAD software of its own.

The required sound absorbing properties required a very heavy fabric construction, a change between three-layer and two-layer technique and, additionally, a non-woven backing, which was bonded to the back of the tapestry. Figure 8.5 shows a woven tapestry after leaving the weaving machine.

Building of the church and development of the tapestries took place simultaneously to meet the deadline for opening the Cathedral in September 2002. Several test hangings of tapestry samples took place on the Cathedral construction site (Nava, 2000). The artist John Nava stayed at the weaving mill for several months to observe the manufacturing process. During the whole project, an intensive transfer of data took place electronically accompanied by two to three visits per year.

8.3.3 Luminescent curtain

There have been a large number of projects aimed at creating luminescence of interior textiles, and many fascinating results have been presented in the past. Examples are the LED-based products of Philips in the Netherlands (Philips, 2010), the neon light wallpapers of Astrid Krogh (Krogh, 2010) or the DigitalDawn installation of loop.pH (Loop, 2010). The following example is a product development that included an interesting attempt to achieve serial production by using automated processes and machines for product manufacture.

The project was initiated in 2007 by research work at Niederrhein University of Applied Sciences focussed on the combination of traditional textile design and electroluminescent elements (Flacke *et al.*, 2009). In a first attempt, the time dependant decrease of illuminating power was taken into consideration. Namely, the lifetime of electroluminescent fibres and cables is limited to approximately five years. Daylight seems to damage the functional layers and a significant reduction of luminosity has been observed by the research team after only one year of daylight exposure.

Therefore, a first version of integrating electroluminescent fibres into a screen printed and fire resistant finished curtain fabric used removable embodiments. Metallic eyes were smashed into the curtain fabric and the flexurally rigid electroluminescent cables were pulled through in order that the illuminants could be removed to wash the curtain fabric and to be replaced from time to time.

Two disadvantages emerged with this method:

- A large amount of time-consuming manual work was necessary for manufacturing.
- Twists of the electroluminescent cables caused the relatively light curtain fabric to bulge locally out of the plane.

Discussions with well-known producers of modern curtains showed that especially the first disadvantage has led to abandonment of similar development projects in the recent past. Weaving of electroluminescent cables into curtain fabrics had been tested but has proved to be inapplicable due to inadmissible damage.

The second version consequently tried to focus on automation of production steps. Cooperation with ZSK, Krefeld, Germany, a manufacturer of embroidery machines, enabled a new method for the integration of electroluminescent cables into a curtain fabric. Basically, a fancy yarn embroidering technology was taken and advanced to process electroluminescent cables. These cables were now fixed permanently to the curtain fabric surface (Fig. 8.6). The design repeats were combined with ordinary, traditional fancy yarns.

With this method, the serial production of luminescent curtains becomes reality. Cost calculation to verify the market potential has shown that dependant on the amount of electroluminescent material for a design, a curtain of three metres length and one metre width causes manufacturing costs of approximately 1000–1500 €.

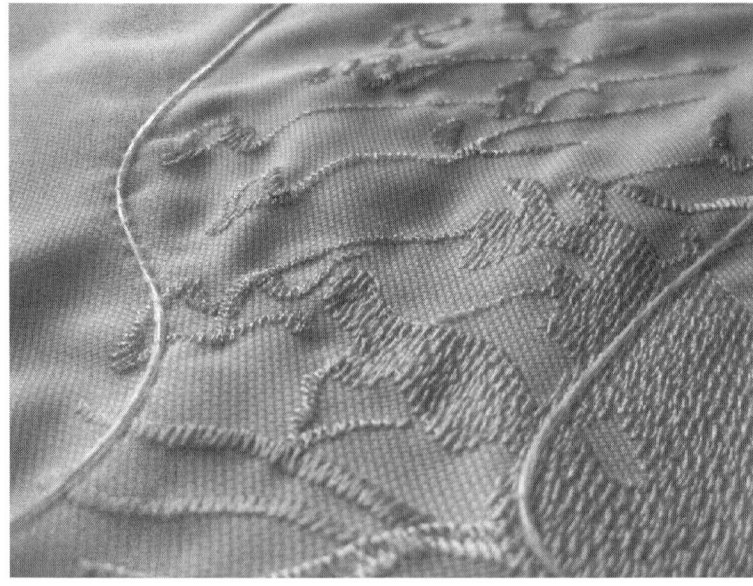

8.6 Curtain fabric with integrated electroluminescent cables by using an embroidering technology (Photo: Eva-Maria Flacke).

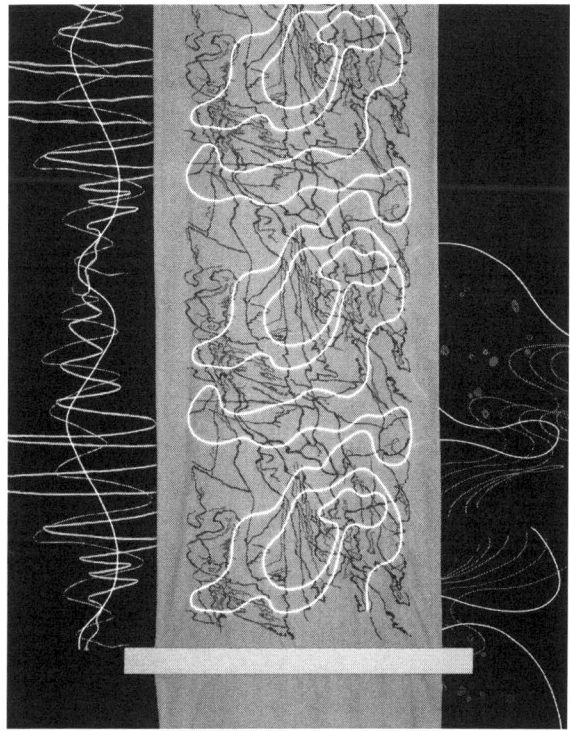

8.7 Illuminated curtain having serial production capabilities (Flacke, 2009) (Photo: Eva-Maria Flacke).

To supply the luminescent curtain with electric power, a new guiding system has been invented and a patent recently applied for. Commercialization is still to be completed. First presentations of embroidery using electroluminescent cables have already been done at machine fairs.

Colour is most important for the design of textiles. Electroluminescent elements add a new main focus to textile design: luminescent colours (Fig. 8.7). This represents an alternative or an addition to dying or printing of fabrics (Flacke *et al.*, 2009). Reproduction ability, automation and cost effectiveness, however, are key success factors for the production of interior textiles. The luminescent curtain project may have contributed to this.

8.3.4   3D woven and smart cover fabrics for automotives

The following example combines two innovative approaches for a new interior textile product. First, a fabric surface should become functional by textile based elements (e.g. to sense a finger touch or to display a recent status or an event by light emission) and second, the basic fabric should be produced seamless by a 3D weaving technology.

Shape weaving is a 3D weaving method originally developed to produce seamless fibre reinforcements for composite parts, getting the required geometry directly by a special weaving procedure (Büsgen, 2008). Several projects transferred this process to interior textile products such as automotive carpets, indoor panel covers or textile based dash boards. A combination of 3D weaving with smart and functional textile elements offers the advantage to manufacture the fabric, create the required shape and integrate functional features with one single step of production (Büsgen, 2009).

The functional specifications are manifold and similar for many projects of smart car interior fabrics. Figure 8.8 demonstrates the general requirements of an innovative dash board cover fabric as an example. A seamless 3D woven cover fabric should contain invisible conductive lines for data transfer and for power supply of functional elements. Light-emitting textile elements had to be integrated to replace plastic enclosed LED's or other conventional light sources. Sensitive surface areas should detect a finger touch to enable, for example, the switching on of a vent, a seat heater or hazard lights (Fig. 8.9). Because cleaning of woven fabric is more difficult than of plastic surfaces, a special finish is required to achieve a dust-repellent, hydrophobic and stain-resistant surface.

A typical challenge in this kind of project is to transfer data between two different CAD software systems. The automotive industry prefers design programs

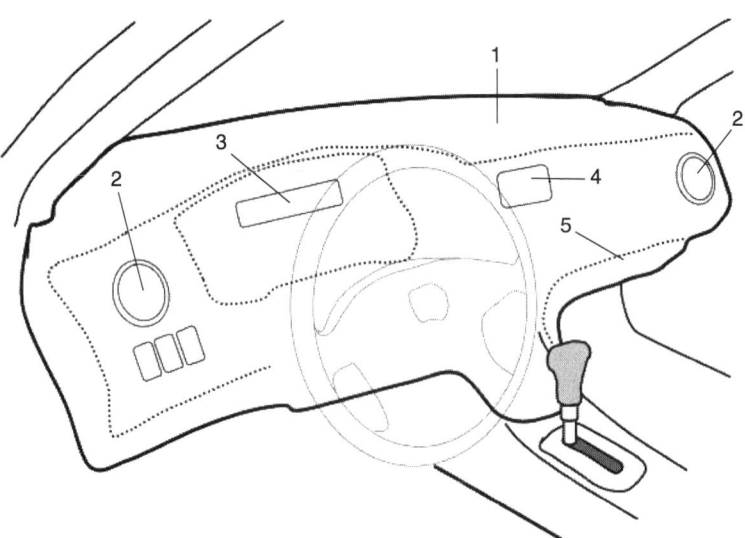

8.8 Basic concept of a 3D woven automotive dash board cover having textile based functional elements (1: woven fabric with integrated conductive threads, 2: illuminated vent, 3: speed indicator, 4: switch panel, 5: luminescent design elements).

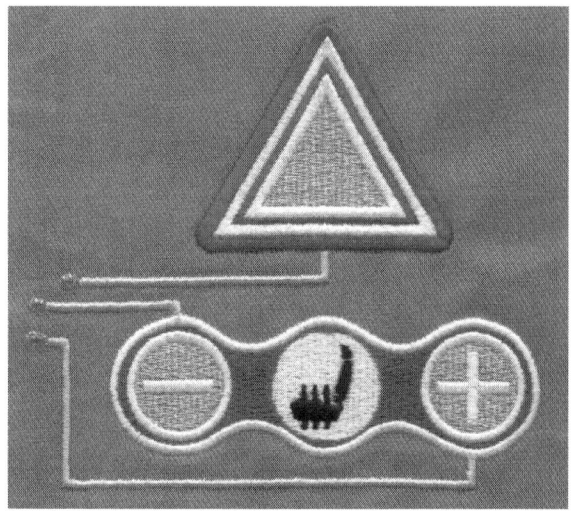

8.9 Embroidered switch panel to control hazard light and seat heating devices.

like CATIA or ICEM. Shape weaving uses a simulation to design and optimize shape and woven fabric construction based on Mechanical Desktop (Birghan *et al.*, 2008a, b, c). Superficially, this kind of transfer is done by standard handover certificates. However, because the basis of software programs is often too different, a design has to be redrawn with the other software and, dependent on the complexity of parts, this can be a time-consuming factor which has to be considered for the project concept.

The next task is to integrate functional elements into a 3D woven surface. Conductive lines and light emitting fibres or cables can be woven in directly, if elements get a separate protection against the intense friction of metallic weaving machine parts such as reed and heddles (Büsgen, 2009) (Fig. 8.10). Switches or keypads may be realized by weaving, as well as adding small 3D bulging areas to the overall 3D shape. Printing and embroidering of switches are possible as well but require additional process steps.

A concept for the finishing of 3D woven geometries involves fixing the flexible fabric by suction of air to a male mould and spraying the finishing material onto the surface.

This project and other similar ones did not enter the commercializing phases until recently. Each development process is nevertheless closing the gap a little bit more. It is open to the future, how far integrated production concepts can play a decisive role for the application of interior textiles in automotives of any kind.

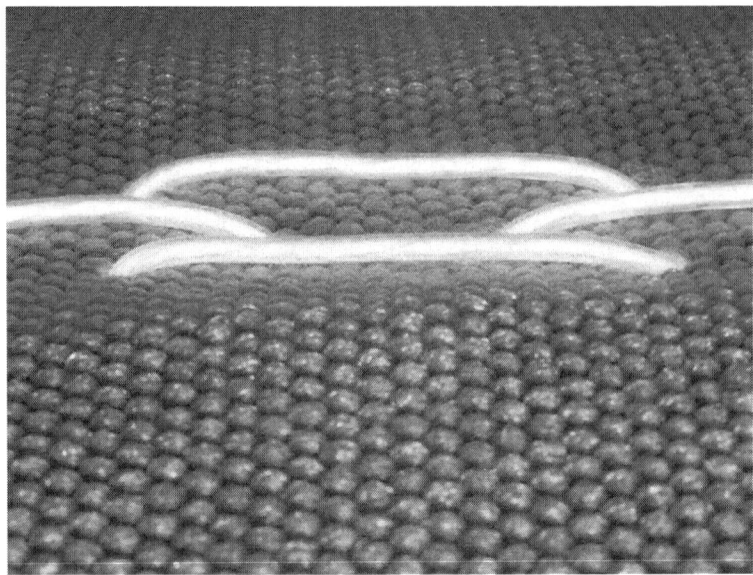

8.10 Light emission by woven in electroluminescent cables.

8.4   Learning experiences for successful new product developments of interior textiles

Interior textiles belong to a long-established market. High quality standards exist and product diversification took place long ago. In fact, the market is characterized by dislocation of production facilities and a takeover of market shares mainly by Asian suppliers. Hence, traditional textile companies in the USA and European countries have switched to speciality products and new developments. The need for those manufacturers to make new products, diversify existing products and remain globally competitive is increasing (el Mogahzy, 2009).

Comparing theoretical procedures of new interior textile product development, as described in Section 8.2, with practical experiences of real development processes shows how far theory deviates from practice in this field. Consequently, only large companies follow a predetermined procedure of product development, whereas small and medium enterprises (SMEs) concentrate on technology and conceptual design aspects. Frequently SMEs commercialize their new product impromptu, using already existing contacts or searching for contacts by chance, for example, through publications or presentations at exhibitions or trade fairs.

It is thought-provoking that developments are aborted too fast, if professional procedures like idea screening and market strategy are taking place or if a scheduled time of development is running out. The first case study in this chapter, about the smart floor project, demonstrates that after the product development left a big company, to be followed up by an SME, the product finally got a chance to enter the market, apparently successfully. A predominant share of new interior textiles is developed by SMEs.

From case studies and other practical product development project experiences the following conclusions can be drawn:

- Following the basic NPD process theories suggested in Section 8.2 helps evaluate the idea and value market chances. However, one single contact (e.g. at a trade fair) may lead to a complete revalidation.
- NPD projects should always include potential end-users to get a clear understanding of their needs and consequently of product requirements. Participating end-users are also often the first customers.
- Systematic conceptual design such as that suggested in Section 8.2 supports decisions about the product concept and may help to justify decisions in post-evaluation. However, it cannot replace creativity and a flash of inspiration.
- Interdisciplinary approaches promise more success for many NPD, if specialized and technically oriented interior textiles should be designed. As a precondition, a company has to expose internal and probably confidential knowledge to project partners.
- Time delay is typical for development processes and needs to be taken into account from the very first moment. Even a delay of years may – sometimes – end in successful commercialization.
- Market reaction can take place much later than scheduled in a plan. The right time and place of commercialization is not easily determined in the first place. Several times a new product has been relaunched two years after finalizing the development work.
- Early publication during NPD (e.g. presentation of a first concept at a symposium) helps to estimate the market reaction. This has already been proved by large companies and should be copied more often by SMEs.

8.5   Future trends in interior textiles

A forecast of future developments of interior textiles should rely on recent research activities and projects in this market. Some trends are selected here, which may lead to commercialized innovative products in future.

Textile walls are coming more and more under the focus of interior designers. Lightweight and easy to transport, these materials offer acoustic and light absorption for public rooms like airport halls or restaurants. Colour changing panels controlled by temperature (Trendhunter, 2008), illuminated fabrics to replace traditional lamps or other additional specifications may support the growth of textile walls in the future.

Phase Change Materials (PCM) are well known for outdoor cloth. Interior textiles can benefit from this material probably more than cloth applications. PCM in screening fabrics, curtains or wallpapers may take up thermal energy during the day and release it during the night, thus saving energy on air conditioning and heating devices.

Multifunctional carpets will play a more important role in the future. SensFloor, discussed in Section 8.3.1, is actually part of a European research project to clarify the application in large crusader ships (BESST, 2009). Interesting ideas like the 'Musical Carpet' of Lim Hyuntaik, Academy Eindhoven/Netherlands (IDSA, 2010), awarded by *Business Week*, may support this trend further.

Leno woven furniture and curtain fabrics represent an example where technology and functionality does not play the leading part. Leno weaving technology has been advanced during recent years, leading to improved products and significantly cost-effective production methods. The product differs from traditional woven fabrics by a clear increase of contrast and brilliance of colours. It may, however, take some time, until the market is realizing and reacting to this innovation.

Until today, smart sensing fabrics only entered the automotive market to a small extent. Because the increase of safety for car occupants is still an unbroken trend, smart fabrics can be expected in this market in the future. One example is seat covers that sense the size and weight of an occupant to adjust the airbag activation and the blow up volume individually. Another promising idea is fabrics that monitor the cognitive status and fitness to drive of a driver.

8.6   Sources of further information and advice

Cathedral of Our Lady of the Angels
555 West Temple Street
Los Angeles, CA 90012-2707, USA
Internet: http://www.olacathedral.org/

Flanders Tapestries b.v.b.a
Rijksweg 416, B-8710 Wielsbeke, Belgium
Tel: +32 56 66 92 44
Fax: +32 56 66 57 78
E-mail: info@flanderstapestries.com
Internet: www.flanders-tapestries.com

Future-Shape GmbH
Christl Lauterbach
Altlaufstrasse 34
85635 Höhenkirchen-Siegertsbrunn
Tel.: +49 (0)8102 89638-0
Fax: +49 (0)8102 89638-99
E-mail: info@future-shape.com
Internet: www.future-shape.com

John Nava: Tapestries from Proposal to Installation
The Cathedral of Our Lady of the Angels Los Angeles, California (editor)
available at Cathedral Gift Shop,
http://www.olacathedralgifts.com/navatapestriesfromproposaltoinstallation.aspx

Shape 3 Innovative Textiltechnik GmbH
Friedrich-Engels-Allee 161
42285 Wuppertal, Germany
Tel.: +49 (0)202 281010
Fax: +49 (0)202 81960
E-mail: info@shape3.com
Internet: www.shape3.com

Niederrhein University of Applied Sciences
Department of Textile and Clothing
Webschulstr. 31
41065 Mönchengladbach, Germany
Tel.: +49 (0)2161 186 6010
Fax: +49 (0)2161 186 6013
Internet: http://www.hs-niederrhein.de/en/fb07/

ZSK Stickmaschinen GmbH
Magdeburger Str. 38-40 · 47800 Krefeld
Tel: +49 (0)2151 444-0
Fax: +49 (0) 2151 444-170
E-mail: zsk@zsk.de
Internet: www.zsk.de

8.7   References

Anon. (2010, April). Executive Summary – Textiles Report – Medical and Sports, http://www.observatorynano.eu/project/document/3122/.
BESST (2009). Breakthrough in European Ship and Shipbuilding Technologies, European Research Project, http://www.besst.it/ [accessed 10 September 2010].
Birghan, A., Tillmanns, A., Finsterbusch, K. and Büsgen, A. (2008a). 'Simulation and calculation of seamless woven 3D shells (Part 1)', *Technical Textiles*, 02/2007, pp. E144f.
Birghan, A., Tillmanns, A., Finsterbusch, K. and Büsgen, A. (2008b). Simulation and calculation of seamless woven 3D shells (Part 2)', *Technical Textiles*, 03/2007, pp. E180f.
Birghan, A., Tillmanns, A., Finsterbusch, K. and Büsgen, A. (2008c). Simulation and calculation of seamless woven 3D shells (Part 3)', *Technical Textiles*, 04/2007, pp. E266f.
Büsgen, A. (2008). 'Simulation and realisation of 3D woven fabrics for automotive applications, 1', World Conference on 3D Fabrics, Manchester, 10–11 April.

Büsgen, A. (2009). '3D textiles – manufacture, applications and possibilities for the integration of functions', NATO – Kiev Conference, advanced research workshop on textile composites, National Technical University of Ukraine Kiev Polytech. Inst. (NTUU-KPI), Kiev/Ukraine, 18–21 May.

Chakraborty, K. (2008). 'Home textiles – global vs. Indian market', Texcellence 08, 3rd National Conference, 3–4 October, Ahmedabad, India.

Federal Ministry of Education and Research (2010). SensFloor – Support and Safety for an Independent Life (booklet).

Flacke, E.-M., Rabe, M. and Oettershagen, A. (2009). 'Elektoluminiszente Elemente in Wohntextilien', *forward textile technologies (ftt)*, June, pp. 74–75.

Flacke, E.-M. (2009). Underwater lab – illuminated pantries, collection concept, Diploma Thesis, Niederrhein University of Applied Sciences, Mönchengladbach, Germany.

Guiltinan, J. P. (1997). *Marketing Management, Strategies and Programs* (6th edn), New York: McGraw-Hill Companies.

Hardt, K. (2006). VTV – Virtuelle Textilveredelung: Schritte zur Simulation von Veredelungseffekten am Computer, Alumni-Symposium, 12 May, University of Applied Sciences, Mönchengladbach, Germany.

IDSA (2010). Sound Carpet, http://www.idsa.org/content/content1/sound-carpet submitted by Faisal Openwave on 23 May 2010 [accessed 10 September 2010].

Infineon (2003). Annual report 2003, Infineon AG, München, http://www.infineon.com/boerse/jahresbericht2003/english/2_3_innovation.htm [accessed 15 August 2010].

Insider (2001/2002), Customer magazine of Lindauer DORNIER GmbH, No. 11, December 2001/January 2002, pp. 6–7.

Kotler, P. (1999). *Principles of Marketing*, Upper Saddle River, NJ: Prentice Hall, pp. 612.

Krogh, A. (2010). http://www.astridkrogh.com [accessed 9 September 2010].

Loop (2010). http://loop.ph/bin/view/Loop/DigitalDawn [accessed 9 September 2010].

El-Mogahzy, Y. E. (2009). *Engineering textiles*. Cambridge: Woodhead Publishing, pp. 3–112.

Nava, J. (2000). http://www.johnnava.com/JNS%20Archive/COS/cos.htm [accessed 31 August 2010].

Nielson, K. J. (2007, July). *Interior Textile: Fabrics, application and historical style*, John Wiley & Sons Inc., p. ix.

Pahl, G., Beitz, W., Feldhusen, J. and Grote, K.-H. (2006). *Engineering Design: A systematic approach* (3rd edn), Berlin: Springer Verlag.

Philips (2010), http://www.lighting.philips.com/gl_en/country/index.php?main=global&parent=4390&id=gl_en_country_sites&lang=en [accessed 9 September 2010].

Rasmussen, J. (2008, May). 'State of the industry 2008, Part I', *Specialty Fabrics Review*, pp. 28–35.

Roman Catholic Archbishop of Los Angeles (2008), Cathedral of Our Lady of the Angels, http://www.olacathedral.org/cathedral/art/tapestries.html [accessed 31 August 2010].

Roth, K. (2000). *Konstruieren mit Konstruktionskatalogen* (2 Vols, 3rd edn), Berlin: Springer Verlag.

Trendhunter (2008). Color changing 3D wall panels – heat-sensitive fabric responds to touch and temperature', *Trendhunter Magazine*, http://www.trendhunter.com/trends/touch-wall-panels [accessed 20 April 2010].

Tromsdorff, V. (1990). *Innovationsmanagement in kleinen und mittleren Unternehmen*, München: Verlag Franz Vahlen, pp. 9.

Trott, P. (2005). *Innovation Management and New Product Development* (3rd edn), Harlow: Prentice Hall, pp. 356.

9

New product development for e-textiles:
experiences from the forefront of a
new industry

P. WILSON and J. TEVEROVSKY, Fabric Works LLC, USA

Abstract: Electronic textiles (e-textiles) are textiles that are, or are part of,
electronic components that create systems capable of sensing, heating, lighting
or transmitting data. Ultimately, e-textiles will have an important role to play in
the fields of medicine, safety and protection. Currently, the industry is still
emerging, and companies interested in this area will improve their chance of
success if they are aware of the challenges they face from technical, business,
regulatory and marketing perspectives. The authors share lessons learned from
the forefront of this new industry.

Key words: electronic textiles, e-textiles, wearable electronics, conductive
textiles.

9.1 Introduction

9.1.1 What are electronic textiles?

The past ten years of development in personal electronics, and changes in
perception of the consumer, have put textiles and electronics on a parallel course
for integration. These trends combined with the economic pressures as a result of
the globalization of the textile industry have made western manufacturers open to
possibilities in textiles that were once only the realm of science fiction. Dubbed
e-textiles, these textiles are, or are part of, electronic components that create
systems capable of sensing, heating, lighting or transmitting data. Currently,
market push is towards novelty products, but product development projects
reaching into military equipment, safety, protection, medicine and beyond have
been and are underway.

9.1.2 Why are they relevant?

The last decade has seen a sea change in the integration of technology in our lives.
The ability to pack ever more computational power and features into a smaller
package has made personal electronics part of our everyday life. As everything
that we carry has a size-to-utility value proposition, the smaller package now
allows us to integrate a larger number of such devices into our daily travels. Once
a unit becomes part of our required set of instruments, fashion and personalization

156

take over, and aesthetics and form become ever more important to individuals. A quick survey of fashion and shelter magazines quickly illustrates that electronic devices have made the leap from *Popular Mechanics* to *Vogue.*

The emphasis on hip design, fashion and wearability of our portable electronics and the aesthetic potential of a device in our decors lead naturally to an integration of the device and the soft goods that surround us. It takes only a cursory examination of magazines such as *Make* or *Craft* to see that the common tinkerer is already fusing these into one to solve problems (Fig. 9.1).

The activity in the craft arena is critically important to the sustainability of the electronic textiles industry. As will be discussed later in this chapter, companies wanting to produce and sell electronic textiles face a number of challenges, one of which being a lack of public awareness of electronic textiles. By embracing electronic textiles, the craft community has preserved the momentum of this emerging industry and shown that the interest in these products exists, while also exposing more of the public to the concept of electronic textiles. The importance of this messaging to the external public is recognized by the industry to the extent

9.1 Several books on handcrafted electronic fashion have been published due to the popularity of the concept among early adopters.

that leaders of this movement are given prominent speaking positions at industry symposiums.

Paralleling the miniaturization and personalization of devices is the coming of age of generations who are more technically literate and use technology as part of their social fabric. Their acceptance of new modes of communication and information sharing is instantaneous. This trait both provides us with a willing young market for e-textile products and one that will be more adept at interfacing with unusual forms for devices. It is a generation that has grown up with light-up shoes, personalized cell phone rings, instant messaging, glow sticks and electronic candy. For them, information should be instantaneous and ubiquitous, and drawing attention to themselves using personal displays of sound and light is socially acceptable.

The last trend that will have a significant impact on the textile industry is advances in 'plastic electronics'. The ability to make conventional silicon-based electronics flexible and durable will help overcome many integration issues that currently plague e-textiles. As will be discussed later in this chapter, textile electronics have limitations in display output and computational density. The ability to substitute polymer light emitting diodes (LEDs) or flexible chip sets for silicon-based ones will marry the best features of silicon devices with the flexibility needed to truly connect well with e-textiles.

9.1.3 What opportunities are there in the marketplace for e-textile products?

Ultimately, e-textiles will have an important role to play in the fields of medicine, safety, and protection. Wearable devices can be made more acceptable with textile-based form factors, increasing the likelihood that the user – whether patient, soldier, policeman or employee – will actually follow through and use them. Personal and portable physiological monitoring, lighting, heating and communications can all benefit from e-textile technology. In the short term, as the industry goes through its growing pains, the individual consumer is the easiest target.

9.2   Integration of electronics and fabrics

Integration of electronics and fabrics can be broken down into three different levels, each of which is very important to the development of this fledgling industry because of the incremental developments they spawn and the markets they establish.

9.2.1 Level 1 – Textiles as platform for outer packaging

At the most basic level, the textile is used as the platform or outer packaging of a conventional electronic device. An example of this type of integration is the

Hoodio™ radio jacket by Wild Planet and GapKids. In this product, the entire radio system is taken out of its tabletop box and distributed around a jacket using the textile as a soft packaging. The jacket provides mechanical stability to the system, but performs no electronic function (Fig. 9.2).

While those committed to e-textiles may cringe to see a fuzzy fabric covering a conventional radio, products such as the Hoodio™ jacket play an important role in creating a model of what an e-textile product is in the mind of the consumer. Acceptance of the convergence between 'soft' and 'hard' and passive and active functions will lead the way to acceptance of more complex integration in the future. This type of product also begins to alter the consumer's value proposition. The Hoodio™ jacket was marketed at $69 – a premium price for a kids' sweatshirt.

The Hoodio™ jacket also revealed some of the challenges that an e-textile product will have to overcome. Some consumers had a knee-jerk negative reaction to having electronics as a part of their clothing – they considered it to be too intimate. In non-apparel products, consumers have expressed safety concerns over plugging in a textile in their home. Knowing these reactions are to be

9.2 Gap Hoodio™ jacket showing distributed hardware using the sweatshirt as a packaging platform.

expected from some consumers allows the product literature, advertising and manuals for e-textile products to address the issues directly, and turn a fear into an educational opportunity.

9.2.2   Level 2 – Portion of the system implemented in textile form

We define the second level of integration as any system in which the textile plays an electrical role in the circuit. This type of integration will dominate the market in the long run and provide the largest market share for textile companies. In this type of integration, the electronics are divided up between the soft and hard in such a way that each is used for its strengths. For silicon chips, it is computational power in a small package. For textiles, it may be softness, flexibility and durability under the environmental conditions the product sees when in use.

Examples of this type of integration are numerous, including the Malden Mills Heat™ blanket, the NuMetrex heart-sensing sports bra by Textronics, Foster-Miller's USB cables, the MET5 jacket by Northface, and the Burton line of iPod compatible jackets and backpacks. Each of these products uses a conductive textile as a sensor, resistor or power and data transmission cable as part of the electronic device (Fig. 9.3).

One particular example of Level 2 integration that showcases how to take advantage of textiles and traditional electronics is the Foster-Miller/ BAE antenna system, developed under a military contract. The antenna was designed to wrap around the torso to be undetectable on a soldier yet operate in a frequency range in which proximity of the human body usually negatively affects the antenna's performance. The actual antenna, also known as a radiator, was implemented in textile form. To make the antenna work in this challenging frequency range, some signal processing had to be done at regular intervals along the radiator. The processing was performed by traditional electronic modules, which were connected to the textile radiators and overmolded with plastic.

9.2.3   Level 3 – Textile based electronic components

The third level of integration and the most challenging of all is the complete use of textiles to form the device. This type of device is farther off in the future, as it requires new textile components and techniques to be invented and developed. Solar cell fibers, battery yarns and transistor fibers are just some of the innovations required. A whole industrial base of fabric circuit design and fabrication technology will be required, as well as a well-trained work force for engineering and fabrication.

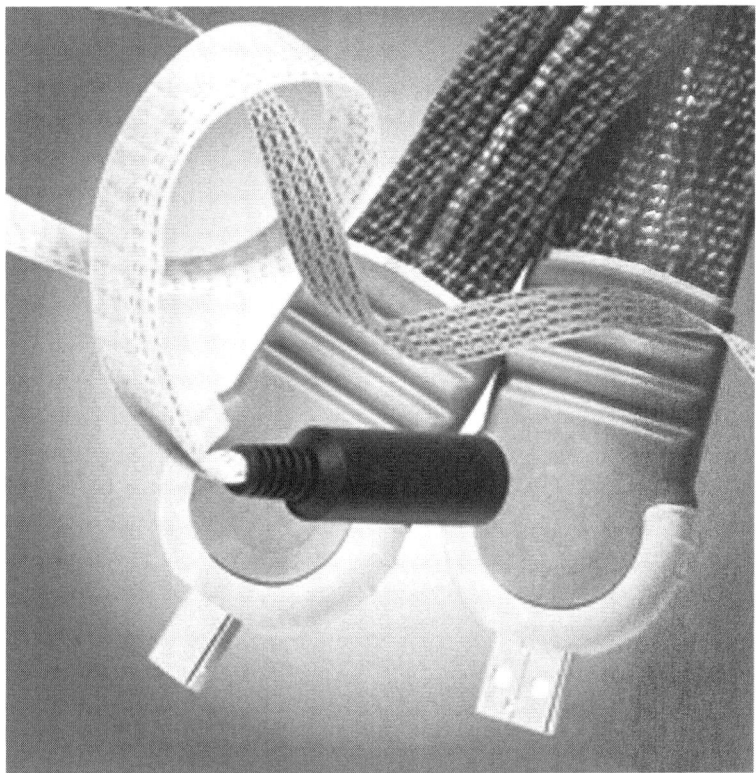

9.3 Textile USB cables developed jointly by Foster-Miller and Offray Specialty Narrow Fabrics.

9.3   E-textiles product development challenges

The contemporary field of electronic textile development is a decade old and, so far, can claim a few small-volume product launches in the consumer sector. Billed as a field with great promise to revolutionize the situation of western textile manufacturers, it has yet to reach this potential. From a global perspective, we can identify several reasons why earlier predictions of multi-billion dollar markets have not yet materialized.

9.3.1   Merging two established industries into one

Every industry has its own specialized language, and the textile and electronics industries may very well top the list of industries with the largest vocabularies. How does an engineer familiar with picks, skew, hand and drape talk to an engineer that wants to know about resistance, wattage and duty cycles?

Once the language barrier is broken, the next hurdle is the expectation of a rapid product development cycle. Currently both industries are very well developed and have short product development cycles. The brands have come to expect a short season cycle and very low dollar investment in new products for their offerings. The need to invest large amounts in long developments to bring an electronic textile product to market is outside of their normal business model and, therefore, takes a large commitment without guaranteed product acceptance.

9.3.2   Impact of the value chain on product development and market success

In any industry there are many levels to the value chain. At the bottom you can see component suppliers from both the textile and electronics sector; above them are first level converters who take raw materials such as fibers, yarns and silicon chips and convert them to broad goods (knits, wovens, etc.) or packaged chips or components. An important segment of the chain is occupied by the integrators, groups who have system knowledge and can take these raw goods and design working prototypes and devices. The integrators then provide these systems to secondary converters, such as garment makers, who work for the brands, which are at the top of the value chain.

When one looks at the slate of companies working heavily in this field, they lie primarily in the textile component and first level textile converters segments or in the brands segment. There is almost a complete absence of electrical component manufacturers, packagers and integrators in the industry. As an industry that wants to accelerate growth and move products to market, we must ask questions as to why these players are absent. Certainly, the exodus of textile production to offshore vendors has sharpened the vision of western textile companies, whereas this transition to local product design with offshore production occurred some time ago in electronics. Therefore, the search for new paradigms is no longer a priority in that industry.

9.3.3   Lack of standards (product, component and safety)

What happens when you make a product in a completely different way than it has ever been made before? You could be making the best product since sliced bread, but you will not get far if you cannot pass the safety testing – not because the product is unsafe, but because in fact it is made in a completely different way and so does not meet the product definition. The Malden Mills Heat™ blanket is an example of this situation. An electric blanket with knitted-in textile heating fibers that runs on low voltage is a very different product from a traditional, thick-wire electric blanket running off of line voltage.

Inherently, the low-voltage product was a leap in safety for a product category plagued by aging issues, such as flex fatigue of heating wires. But, because of known issues for older products, there are well-organized public safety campaigns educating the consumer on the importance of regular inspections and to suspect any product without either a UL or CE mark. To ensure public safety, strict definitions for the product configuration are included in the testing protocol and are based on legacy products. How do you pass UL electric blanket testing if your blanket does not even meet the UL's definition of an electric blanket? To modify these definitions, producers must pay for new standards development, adding time and cost to the effort.

9.3.4   Interfacing standard electronics to textiles

The connector conundrum is a part of every e-textile product development. To really make e-textiles viable in the long run, new packaging types and standards need to be developed for the electronics industry to allow compatibilization to textiles. The volumes of e-textile products on the market have been between 1000 and 20 000 pieces. This is far short of any volume that would justify attention from the component manufacturer and packagers who typically deal in millions of components.

A secondary issue is reluctance by the textile component makers to commit to standards. Unlike solid core conductors, various textile conductors can have dramatically different cross-sectional size to conductivity ratios. The differences arise from the fact that textile fibers are able to be made in a wider variety of constructions – they can be solid cored, plaited or blended to create composite yarns. For standard electrical connectors to be developed, standard sizes for yarns would have to be defined. For a given cross-sectional area, however, some yarns lose in the value proposition; they become too expensive to compete. This fact has made many companies in the industry reluctant to commit to standards, despite the difficulty a lack of standards has caused for all product developments.

9.3.5   Prototyping hurdles

The electronics industry has developed standards that allow all component manufacturers to bring new pieces to market. Developers can quickly prototype devices and manufacture at very low cost using a breadboard system. Such extensive standardization does not exist in the textiles market or the electronic textiles market. Every product development must design each component needed before a device idea can be verified. This lack of an established, standardized breadboard system in e-textiles delays critical go/no-go decisions on product development and increases risk for the companies involved.

Surprisingly, the development of a nascent craft industry around e-textiles has provided some relief for prototyping timeframes. Consistent pressure by fashion students, electronics students, artists and tinkerers on researchers in e-textiles has resulted in the release of developer's kits by companies working in the field.

Concurrently, a National Science Foundation (US) funded project to develop a computational education program targeted at young girls resulted in the development of a set of sewable electronic printed circuit boards (PCBs). The Arduino Lily Pad designed by Professor Leah Buchley of MIT Media Lab is based on the open-source Arduino test boards and takes advantage of a wealth of pre-existing web-based instruction (Fig. 9.4). These recent additions to the toolkit of researchers will reduce times for product development and spur new product ideas from fashion students, artists and crafters which can be monitored and appropriated by textile companies.

9.3.6   The search for a killer app

Every nascent industry is filled with companies looking for the killer app, and e-textiles is no exception. The particular challenge in the field of e-textiles is that consumers lack a mental model for the types of products that e-textiles typically produce. As such, consumers do not understand what they are seeing when they encounter an e-textile. Oddly enough, even highly technical people will succumb to the 'must be magic' mindset when faced with a fabric heater. Thus, the search for the killer app must go hand-in-hand with a program of marketing and education about e-textiles in general, not just the individual product itself. Until the average person is aware that you can make a fabric heater, market pull for e-textiles will

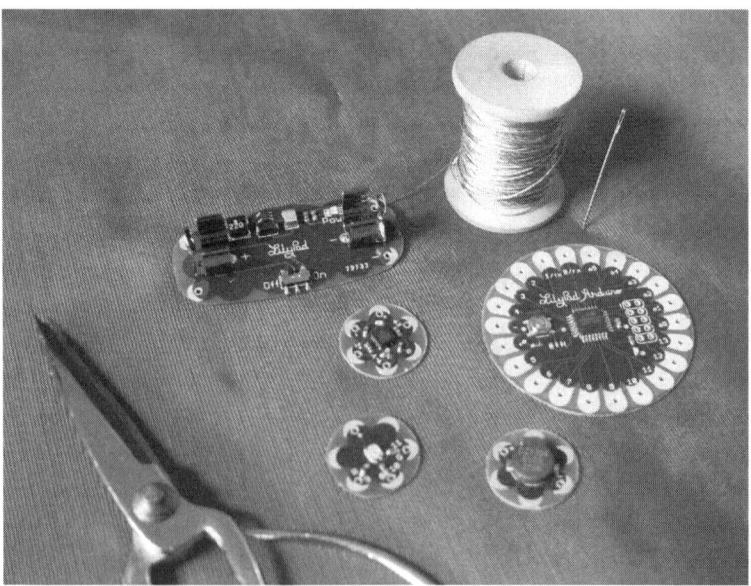

9.4 Lily Pad Arduino sewable microcontroller and accessory power, input and output modules.

be negligible, and the burden of finding the killer app will rest in the hands of the few e-textile experts.

9.4   What every company should know

9.4.1   You will become an electronics expert

The lack of strong integrators in the value chain is forcing most textile companies interested in the e-textiles market to wear more hats in more fields of expertise than they were previously. The development of the industry is not yet at the point where yard goods of cable components or circuits can be sold to a converter. In most situations you will need to design the finished e-textile product yourself and teach the brands how to integrate the device into their products. Because of this, you will need more knowledge on the electrical side than you may have originally thought – in fact, you will need at least one member of your team to be 'bilingual'.

The first area to understand is what portions of an electrical device a textile can replicate. A typical device is composed of:

- an input such as a switch or a sensor
- a computational circuit
- transmission paths between components
- an output such as sound generation, LEDs, display or transmission antenna
- a power source
- packaging.

Areas where textiles can substitute well are in the input, transmission paths, packaging and some limited pieces of the output devices (i.e. antennas). The computational units are best performed by conventional electronics, which have an information density advantage. And at this current time, the output devices and power sources have not yet been well developed in textiles, so conventional pieces are required.

Because most systems will be a hybrid between standard electrical parts and textiles, three areas constantly limit our designs: power, packaging and connections. The size and capacity of the power component, usually a battery, often drive the application and its form factor. Additionally, the need to replace the battery can affect the cost of the product if rechargeable versus throwaways are being considered. It is necessary to understand the battery media (gel, solid state, etc.) and what environmental conditions will do to that battery. Coin cells, for instance, explode in the dryer.

Packaging of the electrical components against mechanical wear and flexure and environmental ingress of dirt and water is important. Textile products are more flexible and durable against rough handling and wet conditions than electrical components (just try to send a conventional wire harness through the washing machine). Breaking of connections and water-induced shorts are common when adding electronics into the textile environment. Proper sealing and stress relieving of components and connections is a big task for a product development.

The last and most important area is connections. The difference in size scale, flexure, material type and form factor makes connections between textiles and electric components very difficult. In the textile environment physical connection techniques that have been developed for textiles are often found to be the most durable for making electrical connections also. These include snaps, grommets, knots and seams or tack bars (Figs 9.5 and 9.6). Often it is the connector from the electronics industry that breaks first.

Product developments to date have for the large part focused on single use, optimized textile parts. To gain the most leverage from your development investment, e-textile components that can be used in multiple product lines will be worth the extra time and design work. A good example of a leveraged design is the Cool Blue™ system marketed by Lands End. An electroluminescent unit was integrated into a suite of products that includes jackets, dog covers, briefcases and backpacks (Fig. 9.7). Reflective tape is used to augment the smaller lighted unit on large area items instead of designing larger lighting units. Creative engineering will be required to design goods that can be produced by the yard and used in multiple electrical components and products.

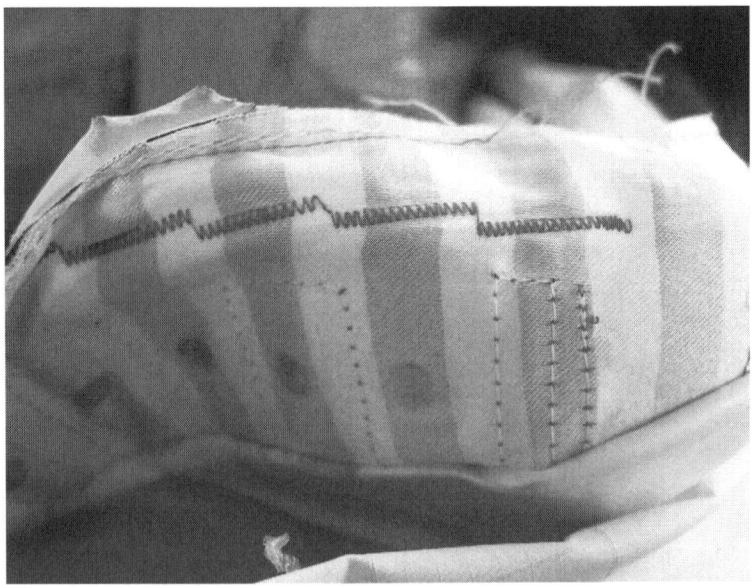

9.5 Tack bars as connectors for e-textile garments.

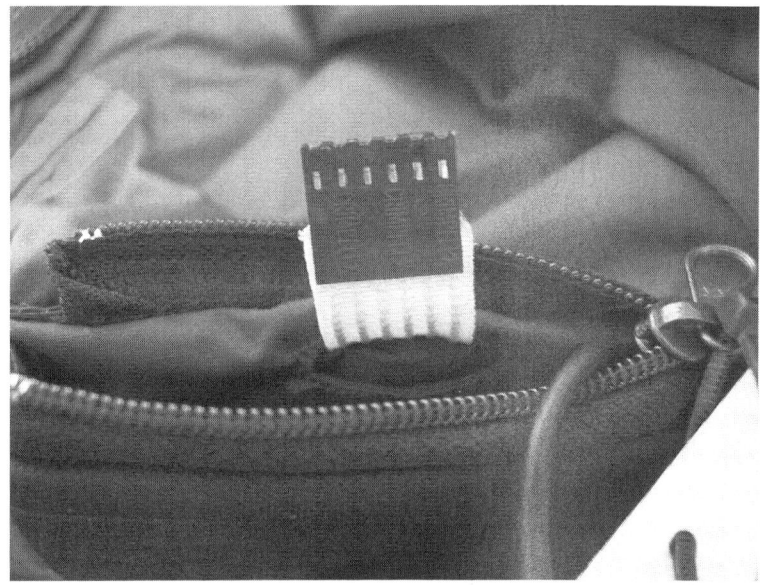

9.6 Standard insulation displacement connector in an e-textile garment.

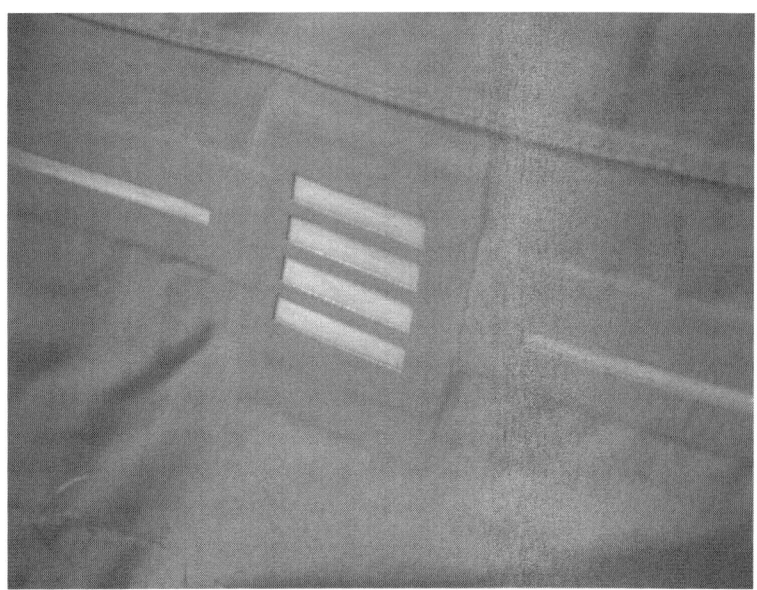

9.7 Close up of a Cool Blue child's jacket showing the two types of elements repeated through all Cool Blue products: a square electroluminescent polymer lamp and a strip version. Cuts in the overlay fabric show portions of the lamps.

9.4.2 New standards and regulations to follow

The regulatory environment is one that must be investigated at the outset of a product plan as the time frame to comply can approach more than a year in some cases. Four US agencies or testing organizations that may impact developments are the Federal Communications Commission (FCC), the Federal Drug Administration (FDA), and Underwriters Laboratory (UL) or ETL SEMKO. If the product will be sold in Europe, the European equilivants to these will need to be contacted (CE and RoHS, for example.)

The FCC regulates the use of radio communication bands and the interference of electrical devices with devices transmitting and receiving. Many e-textile developments need remote controls or use wireless transmission on the product. Poorly shielded and grounded cables in some transmission systems can become antennas themselves, changing the frequency band where they are transmitting. Part 15 of the code contains the regulations for devices that are intentional or unintentional radiators. These regulations spell out the radiation power as a function of distance from the device that is acceptable in each of the communication bands. The limited number of bands used for transmission and reception (900 MHz, for example) results in several very crowded spaces and many unintentional interferences, as those of us with cordless handsets at home can attest to. If your device is covered by Section 15, you will need to have your product submitted to an established testing house for verification, certification or declaration of conformity to FCC regulations. In short, verification is for unintentional radiators (usually with high frequency vibrations), certification is for those that intentionally radiate, and certification covers most computational equipment. This process can take anything between a few weeks and several months depending on what type of approval is needed. At first glance, product developers in e-textiles might dismiss the need for their product to be examined under these testing regulations due to naivety about electronic devices. But in fact, there have been circumstances where electric blankets have been caught as unintentional radiators resulting in large fines imposed upon the brand and manufacturers for importing them without FCC verification. The FCC has an army of inspectors who track down complaints of television, computer and handset noise.

The FDA can be a complicated agency to deal with. In short, the FDA will be a concern depending on how you use medically related data acquired by any monitoring system that you develop. At the far end of the spectrum, if diagnosis will be made from the data (i.e. alive or dead, shock, etc.), then the entire garment could become a FDA regulated medical device and have to be submitted to the 510K approval process. On the other end, lifestyle products (those that give you feedback on your heartbeat, temperature, etc.) can fall under privacy regulations if the medical data is transmitted or stored by a second party other than the person using the device.

A lesser known need for FDA compliance is biocompatibility testing for a device. Depending on the level of product regulation, the effect of the materials in the design on biological activity might require testing. One simple example is the use of a physiological monitoring device in a clinical research setting, a typical first step for companies entering this space. For such a device to be used, the Independent Review Board (IRB) overseeing human test subjects would require that your device comply with ISO-10993 Biocompatibility tests. These tests even apply to skin contact with fabrics. Product designers will find very limited biocompatibility data exists in the textile industry outside of specific medical textiles and will be forced to run tests on the component's materials. Some e-textile fabrics with silver coatings can have adverse affects on cell cultures in the testing and this must be considered early in product design.

Underwriters Laboratory (UL) and ETL SEMKO are nationally recognized testing laboratories that give a stamp of approval to electrical products sold in the consumer marketplace. Their stamp signifies that a product has been tested in accordance with safety regulations developed by standards groups and has passed. Accompanying this is a requirement by the manufacturer to allow periodic inspections of its manufacturing by inspectors to ensure that the product is being fabricated to specification. Over 900 categories of devices can receive UL or ETL approval. While not required by law to introduce a product to market, the consumer awareness of UL stamps in some categories is high and might affect how the product sells. Certainly it is a consideration in your risk management plan for product liability. Unfortunately, most standards are a combination of product specifications and performance standards. In most cases, any product which substitutes a textile material for an electrical component will automatically be unable to meet the specification portion of the evaluation and may not be able to receive a mark until they develop a new specification for your product type, including modified test procedures. This will likely be at your cost and have a longer time frame associated. Therefore, knowledge of your options is required in your market plan.

In certain situations subcomponents of your product (i.e. transformers) will be able to receive a mark and can help to mitigate consumer perception. Which testing laboratory is used is a combination of time frame and which mark is more recognized by the target consumer.

Other areas where third party approval may be needed are for sales to the military and those using data transmission protocols. The military is slowly moving away from military specifications to performance standards that will improve the inclusion of e-textiles. As for data transmission protocols, our earlier work with cables found that while we could design a textile cable that could meet the performance specifications of standards such as USB, we could not call it a USB cable with the markings unless the standards were changed by IEEE, a long process that is still not complete eight years after first being raised.

9.4.3   Manufacturing for product reliability and safety

Careful product testing using an established testing house can help avoid problems after the product has reached the market. There have been several recalls of products with a mixture of textiles and electronics. The latest of these is the Cool Blue™ backpack. Insufficient heat dissipation around the battery unit can cause excessive heat buildup and softening of the plastic housing. Textiles are a natural insulator and the proper thermal management of an electrical system requires some thought when designing the product. Note that computers all have fans to eliminate the heat generated by the large number of computations through even low resistance copper.

Proper controls during manufacturing can also eliminate future problems. It is important to become familiar with recommended equipment for making connections, testing protocols, electrical breakdown testing, circuit testing, etc. during manufacture. Better yet, establish a manufacturing partner whose business is electrical devices and is experienced in manufacturing safe devices. A significant barrier for this industry will be the establishment of manufacturing centers where low-cost textile fabrication (garment, homegoods, etc.) and assembly-line electrical device manufacture can co-exist. Sewing over an insulated wire or using an untrained garment worker to make an electrical connection can lead to high reject rates or problems in performance or safety later. For textile companies, working with a systems integrator who can warranty either the entire system or the electrical components is a way to reduce manufacturing risk.

9.4.4   Import and export overseas

Moving electronic products around the globe is much more complicated than moving textile products. If your product contains a battery, you will likely encounter restrictions on shipping via air. Coin cell lithium batteries, for example, cannot be shipped via air in any sizable quantities, and may require special testing and documentation. Lithium ion batteries are easier to ship by air, but still have quantity restrictions on them. Therefore, if you are manufacturing your product in Asia and planning to sell it in the US, the lowest risk approach is to plan container shipping by sea into your timeline.

Clearing customs is also more complicated with e-textile products than with textile products. The product codes used to assess the duty rate may not fit your product exactly. Your electronic design partner should be able to help you select the code that best fits your system.

9.4.5   Regional manufacturing bases and integration of product components

If you are manufacturing in China, your electronic and textile components may be made in different provinces. Surprisingly, moving components from one province to another for integration within China may prove to be more difficult than exporting the components into Hong Kong, for example, then re-importing them into the province that will be integrating the product. Another option is to import the separate components to a third country for integration. Importing components into the country you are planning to sell in and integrating there may cut down on shipping costs, but will add to labor costs for integration. Since shipping costs are non-trivial, your logistics trail needs to be thought through as soon as possible and built into the cost of the product.

9.4.6 Designing for the buyers

A subtler barrier to entry for a product is understanding the money trail. Outside of the consumer marketplace where the buyer is usually the user, a purchasing infrastructure has likely been established to use company or municipality funds to provide equipment. A fabulous product that the users have been waiting for may be obvious to them and us, but combining multiple functions together may make the product inappropriate for the purchasing infrastructure. Some markets are dominated by a few large municipalities (e.g. NYC or Los Angeles), which must be convinced to use the new hybrid product before the rest of the market will bite. There also may be different offices and standards for electronic equipment versus protective clothing. Getting both of these buyers to agree on a system may be a challenge, not mentioning how their accountants will deal with the budget line items. Simple design changes that allow modularity or different part numbers may get your product around these challenges.

Before embarking on a product development, do some ground work to find out if there are institutional hindrances to your product which you can treat in your product design. If equipment is often resold by the original buyer when obsolete in their district, build more detachable components into the system to allow broken parts to be replaced or for the entire system to be removed. In the safety and protective fabrics industry, chances are that the systems you are developing are expensive and will outlast the electrical components. Therefore, plan for component replacement or graceful degradation of the product. Ensure that, if the electrical portion of the device fails, the rest of the system can perform to some minimum useful level. While full integration may be the 'holy grail' for elegance of design, it might not be the most practical solution and will be immediately noticed by buyers.

9.4.7 Duct tape opportunities – finding new product ideas

Look for what we like to call 'duct tape opportunities'. In the world of superusers of equipment, those who are actually doing the jobs know what they need better than those that sit in the laboratories. A few hours accompanying your consumers and observing their equipment and use patterns will give you many product ideas.

It did not take the engineers at Carnegie Mellon University long to watch aircraft mechanics go back and forth to a truck full of manuals to realize that a wearable computer would speed repairs. And how long did it take product developers to realize that electrical tape was being used to secure excess webbing on load carriage equipment before they started supplying straps with elastic to perform the same function? People can be very innovative with a little duct tape when they need to be hands free. These jury-rigged systems are all products waiting to happen.

9.4.8 Controlling the marketing

If the product development team has done a good job, your e-textile product will not look that different from a traditional version of the product. For example, a light-up jacket will look pretty much like any other jacket when it is hanging on the rack. Good integration is important for the consumer to accept using the product – and creates an enormous challenge in marketing the product. If the technology is invisible to the consumer, how will he/she be able to differentiate your e-textile product from its inactive counterpart? Since price points for textile products vary so widely, consumers can easily fail to realize that the jacket with the higher price tag has an active functionality. One case in point is the launch of a series of private label Champion jackets for boys at Target. The integrated electroluminescent lighting by SafeLite in decorative patches on the shoulders was meant to provide safety illumination for children at night. At the same time Target retailed a lower cost, modified version of the same jacket with very minor visual changes. Hanging on the same rack, the jackets were virtually indistinguishable to the consumer (Fig. 9.8). The point of purchase signage was not effective in bringing the uniqueness of the product to the attention of the consumer. SafeLite had anticipated the need for illuminated POP signage and had developed a program using their expertise in inexpensive lighted signs. Unfortunately, the driving forces to deploy provided POP were not as well understood and compliance at each Target was low, resulting in difficulty in reaching consumers.

Most e-textile products are produced at low volumes initially, and come with small promotion budgets. Getting the visibility to effectively market your product can be very difficult. One frequently used channel to market is catalog shopping, because it is relatively easy and inexpensive. The difficulty with an effective catalog listing, however, relates back to the lack of a mental model in the mind of the consumer and the invisibility of the technology in the product. A picture of the product is not enough – some sort of eye-catching icon or symbol that says 'this product is not what you think it is' is highly recommended. A catalog does offer the opportunity to explain the product, but only if you have managed to catch the attention of the consumer in the first place.

(a)   (b)

9.8 SafeLite Champion jacket with side arm patch racked with non-active jackets of similar design and Target point-of-purchase marketing.

9.5   Sources of further information and advice on e-textiles

As an industry, e-textiles crosses multiple boundaries and, therefore, has not found a well-established industrial organization to call home for conferences and journal articles. There are several venues to monitor to find the latest information and to network.

Conferences include the targeted Smart Textiles conference hosted by Pira International, usually held in the spring. Pira also holds sections on e-textiles during related fields such as plastic or organic electronics forums. The Safety and Protective Fabrics division of the Industrial Fabrics Association International (IFAI) periodically hosts specialized sessions treating e-textiles. Techtextil in Europe offers a place for papers and exhibitors to show off their latest product developments. A formerly popular venue is the International Symposium on Wearable Computing (ISWC), but this has transformed into more of an academic event and is less frequented by corporations. There are several technical volumes that have appeared as compilations of technical papers; these can be found in the references section. A number of general reviews of the field or historical developments have been published as part of larger volumes on technical textiles. The most notable of these are *Extreme Textiles* and *Technical Textiles 2*.

An interesting genre of texts is those written for the handcrafter as how-to volumes for making personal e-textiles. A number of these were published recently and should be consulted to understand both how the craft field is impacting the industry and to review creative product concepts. The authors of these books host active websites and blogs that serve as a means for others to post their projects.

The most influential website for e-textile professionals is www.talk2myshirt. com. Written by an industry insider, the website is a clearinghouse of information on new products, trends and materials.

9.6   Conclusions

A quick review of the web can illustrate how e-textiles have captured the imagination of early adopters and influential thinkers while products to dominate the market have been waiting to appear. The potential in fields such as medicine, safety and protective equipment and performance clothing is obvious. But to reach this potential, companies who desire to fill voids with products must be aware of the issues that appear when the development straddles two disparate fields. Knowing the potential pitfalls of product development and anticipating them can enable more products to successfully get to market on time and in a form that is engaging to the consumer and meets their needs and expectations.

9.7   References

Braddock Clark, S. (2005). *Techno Textiles 2: Revolutionary fabrics for fashion and design*, London: Thames & Hudson.

Eng, D. (2009). *Fashion Geek: Clothes. Accessories*. Tech, Cincinnati, OH: North Light Books.

Marzano, S., Green, J., van Heerden, C. and Mama, J. (2000). *New Nomads: An exploration of wearable electronics by Philips*, Rotterdam: 010 Publishers.

McCann, J. and Bryson, D. (2009) *Smart Clothes and Wearable Technology*, Cambridge: Woodhead Publishing.

McQuaid, M. (2005). *Extreme Textiles: Designing for high performance*. Princeton, NJ: Princeton Architectural.

Pakhchyan, S. (2008). *Fashioning Technology: A DIY intro to smart crafting*, Sebastapol, CA: Make Books.

Switch Craft (2008). *Battery-Powered Crafts to Make and Sew*, New York: Potter Craft.

Talk2myShirt, http://www.talk2myshirt.com [accessed 30 September 2009].

10

Customer co-creation: moving beyond market research to reduce the risk in new product development

F. T. PILLER and E. LINDGENS, RWTH Aachen University, Germany

Abstract: Forecasting the demand for new products is becoming increasingly difficult in many markets. A new method to decrease the failure rate of new products builds on the idea to integrate customers deeply into the innovation process. This process of 'customer co-creation' starts with an open call for participation to a community of designers and hobbyists to submit a new product concept. Reactions and evaluations of other consumers towards the posted concepts are encouraged in the form of internet forums and opinion polls. The product concept is then presented to the customer community for evaluation. Only if the number of interested consumers exceeds a minimum threshold, are investments in final product development made, merchandising settled, and sales commenced. This paper illustrates this process of customer co-creation with the example of Threadless.com, a pioneering company in the apparel market from the US.

Key words: Customer co-creation, new product development, collective customer commitment, forecast.

10.1 Introduction

The manufacturer's nirvana is to develop and produce exactly what its customers want, when they want it – ideally with no risk of overstocks or inventory. The increasing heterogeneity of demand, a rapid change of preferences, and the resulting micro-segmentation of many product categories, however, prevent manufacturers from reaching this state easily. In many consumer goods markets, manufacturers today are forced to create suitable assortments for smaller market niches than ever before, as these markets frequently are the only way for growth and an escape from heavy price competition. In such a situation, new product development projects often cause enormous investments and are highly risky. While new products or product variants have to be developed and introduced at high pace, forecasting their exact specification and potential sales volumes is becoming more difficult than ever.

The purpose of this article is to illustrate a new approach to address this challenge. Its basic idea is not to ask customers for feedback on decisions made by the company, but on integrating the customers deeply into the processes of the manufacturer or retailer. We will introduce the idea of 'co-creating with customers'

175

and present an in-depth case study of a pioneering company in the industry, Threadless.com, that has implemented co-creation with great success. Building on this case study, we will suggest further examples of how and when to apply customer co-creation in the textile and apparel industries.

10.2　Challenges of identifying customer needs in the product development process

Recent research studies confirm large failure rates in new product commercialization.[1] Newly launched products have shown notoriously high failure rates over the years, often reaching 50% or more. The primary reason for these flops has been found to be inaccurate understanding of user needs. Many new product development projects are unsuccessful because of poor commercial prospects rather than due to technical problems. Research found that timely and reliable information on customer preferences and requirements is the most critical information for successful product development.[2] Conventionally, heavy investments in market research are seen as the only measure to access this information. The apparel industry in particular, with its fast changing trends and collections, is facing huge challenges. Companies such as H&M or Zara have set the pace with their 'fast fashion' concepts. Following and copying these models has not been profitable for all companies.

So the basic question remains: how can a company identify the customers' needs perfectly in order to forecast their future desires, and design and produce optimal assortment on this basis? One opportunity to handle these challenges is shown by an extraordinary company called Threadless (threadless.com).[3] Besides reducing inventories, eliminating markdowns and increasing customer loyalty, they do a marvelous thing, that is, produce exactly what their customers want by co-creating in the product development process. How is this different from a conventional company? Most fashion companies also 'ask' their customers what they want by various means of market research. But the difference in the Threadless approach is that (i) they ask not just a sample, but almost every consumer of their products, (ii) they test every single product concept, and (iii) the decision about their assortment composition is entirely based on the customer's feedback.

Before analyzing the Threadless model in more detail it is helpful to briefly review the conventional model of creating a new product assortment for a fashion company. Common wisdom says that to learn about customer preferences and requirements, companies should invest in market research activities. Questionnaires, surveys, or interviews ask consumers what they like and dislike. Among the methods for testing new concepts, the most common are focus groups. They are popular because the results are easy to interpret and the method is fast, inexpensive, flexible and confidential. Unfortunately, focus group research has a number of severe limitations.[4] One problem is that the results from a test with a few consumers are not a reliable indicator of the reactions of the broader population. In addition, focus groups lack realism. Consumers have to react to verbal descriptions of concepts or a rendering of a product. As a consequence, this research method tends to underestimate the benefits of a truly unique new product concept. Focus group research – as most other common market research methods – does not measure real consumer purchasing behavior. It reveals information about the consumers' attitudes toward new products or their intentions to purchase them. But it does not provide quantitative estimates of sales, market share, product cannibalization and profitability. More reliable and accurate measures, such as test markets, demand expensive set-ups and take a very long time to deliver results. Finally, there is a high level of noise in these tests, such as competitors' activities, manufacturers' advertising and economic change. Most market research measures demand background data to calibrate forecasting or to correct for biases in stated purchase intentions.

In anticipation of these problems many companies perform no market research at all. Studies of the actual practice of market research report that companies regularly fail to undertake thorough market research and use only very few of the available tools and methods to include customer input in the development process. A survey of Fortune 500 firms found that only the focus groups method was used by more than half of the companies studied, and only two other methods (limited rollouts and concept tests) were used by more than 25% of the respondents.[5] This is rather surprising, given the huge amount of scholarly study and a whole industry providing these market research services. One frequent excuse is that customers are difficult to predict: they often cannot express what they want or are internally inconsistent, often many people with different needs are involved in one purchase decision, and it is likely that customers have changed their minds by the time the product is launched. As a result, many manufacturers tend to stick with existing assortments, building their new products first of all on a revision of the existing offerings. This may improve the capability to forecast demand for new variants, but places suppliers in persistent danger of missing important trends. It also prevents them surprising their customers with really new products and innovative applications.

10.2.1 Threadless.com's idea to substitute market research expenditures by sales

Threadless, a young Chicago-based fashion company, follows an innovative business model that allows it to create a high variety of products without risk and without heavy investments in market research to access customer preferences before production starts. In fact, it follows a strategy that turns market research expenditures into quick sales. Its business model has been called 'the most innovative start-up' in 2008.

Started in 2000 by designers Jake Nickell and Jacob DeHart, Threadless focuses on a hot fashion item, t-shirts with colorful graphics. This is a typical hit-or-miss product. Its success is defined by fast changing trends, peer recognition and finding the right distribution outlets for specific designs. Despite these challenges, none of the company's many product variants ever flopped. But Threadless has neither sophisticated market research or forecasting capabilities, nor a complicated flexible manufacturing system. Rather, all products sold by Threadless are created, inspected, improved, approved and selected by a user community before any larger investment is made into a new product. Together with 51 employees, the company's founders sell more than 150–160 thousand t-shirts per month for between $18 and $24 each with a 30% profit margin on sales. Sales in 2009 hit almost $30 million, with profits of roughly $9 million. Since 2006 annual growth continued at more than 150%, with similar margins. Threadless has 1.5 million followers on Twitter and more than 100 000 fans on Facebook. The company's website has logged 2.5 million unique visitors in August 2011, a 50% increase over the same month last year. This was achieved by transferring all essential productive tasks to their customers who, in turn, fulfilled their part with great enthusiasm. Customers design their own t-shirts and help improve on the ideas of their peers. They screen and evaluate potential designs, selecting only those that should go into production. Since customers (morally) commit themselves to purchase a favored design before it goes into production, they take over market risk as well. Customers assume responsibility for advertising, supply models and photographers for catalogues, and solicit new customers.

Threadless is a textbook example of customer co-creation. In the literature on innovation management, the active role of customers in the development process has been studied in the last few decades. In this domain, the term co-creation (also co-design, user innovation or open innovation with customers) denotes a product development approach where customers are actively involved and take part in the design of their own product. More specifically, co-creation has been defined as an active, creative and social process, based on collaboration between producers (retailers) and users to generate value for customers.[6] Customers are actively involved and take part in the design of their own product and their co-creation activities are performed in an act of company-to-customer interaction and cooperation.[7] The method breaks with the known practices of new product development. It utilizes the capabilities of customers and users for the innovation process. Research has shown that many commercially important products or processes are initially thought of by innovative users rather than by manufactures. Especially when markets are fast-paced or turbulent, so called lead users face specific needs ahead of the general market participants. Cooperating with them has been described as an important source of innovation for firms. The main benefit of customer co-creation is to enlarge the base of information that can be utilized for the innovation process. Need and solution information of the firm is extended by the large base of information about needs, applications and solution technologies that resides in the domain of the customers and users of a product or service. Customer co-creation supplements a firm's internal innovation activities, but does not substitute them.[8]

So how is co-creation working at Threadless? The process starts when an idea for a product is posted on a dedicated web site by a community member. All new designs are submitted entirely by the community, which includes hobbyists, but also professional graphic designers. The company exploits a large pool of talent and ideas to get new designs (much larger than it could afford if the design process would have been internalized). Creators of submissions that are selected by other users get a $2000 reward, $500 worth of free t-shirts, and their name is printed on the particular t-shirt's label. In fact, Threadless has over one million registered users and receives approximately 300 submissions per day.

Next, the reactions and evaluations of other consumers towards the posted idea are encouraged in the form of internet forums (comments) and opinion polls. Users evaluate new designs on a scale from 0 ('I don't like this design') to 5 ('I love this design'). On average, each design is scored by 2500 people, and about 90–100 write an explicit comment on each design. A good score corresponds to a value above 3.0. In addition, customers not only express their marked preference for specific designs, but can also opt-in to purchase the design directly once it has been chosen by the collective. For this, they check the box 'I'd buy it' next to the scale, providing an informal commitment to later purchase the product concept if it is selected. From the designs receiving the top votes and largest commitment of users to purchase, Threadless is currently producing between four and six new products each week. New designs regularly sell out fast, but are reproduced only if a large enough number of additional customers commit to purchase a reprint.

This process has been called 'collective customer commitment'.[9] It exploits the commitment of users to screen, evaluate and score new designs as a powerful mechanism to reduce flops of new products. In this way, market research expenditures are turned into early sales. The origins of the idea can be traced back to Kohei Nishiyama and Yosuke Masumoto, two industrial designers from Tokyo. In the 1990s, they pioneered the idea with their company Elephant Design. The core element of the company is its website cuusoo.com (*cuusoo* means 'ideal' or 'daydream' in Japanese). Here consumers can post ideas for desired products. One idea, for example, came from a copy-editor who used his home as an office and wanted a discreet microwave, a plain white box. This seems to be an odd request, but when the company showed a virtual prototype, many users expressed interest. After a sufficient number of consumers had expressed their commitment for this kind of microwave (in this case not just by clicking a box as in the case of Threadless, but by even making a small down-payment of about $19), the company decided to invest in further product development and finally launched the product to the market.

It is important to note that in the end management has the final word. The Threadless team reviews each short-listed design to make sure that no user cheated by analyzing IP addresses and IP chains for voters and the respective scores given. But more fundamentally, Threadless' team also has its own say in the selection process. The company learned that the collective input of their customers has to be combined with the companies' internal market knowledge to succeed with the commercialization of the selected products. At Threadless, the winning designs are chosen from among the top scoring designs, but they are not necessarily the highest scoring designs. Important factors are the originality of the design (is it somehow timeless, not too similar to other recent winners), legal issues (are there any copyright related issues), and assortment policy (will the design contribute to a wide assortment of products). For this decision process, however, the community again provides important input. The often long list of user comments about each design provides helpful information if a design is plagiarized, but also if it could be modified to look better.

Over time, Threadless has refined its customer co-creation process. For example, to keep the competition interesting and encourage users to participate continuously, the number of designs at any one given time has to be limited so that users do not get confused. Usually, each design gets seven days to be scored. But if a new design has received a low arbitrary score (made up of multiple variables, including the number of 'I'd buy it' requests and the design's average score) within the first 24 hours of its positing, it will be dropped from the running. This happens to about one third of the submissions. The early user feedback has proven to be a very strong indicator of the success of a design in the competition and enables the company to increase the usability and experience for users who vote. It is important to know that Threadless is focusing customer co-creation on the product features where uncertainty is largest – the aesthetic design. The elements where the fashion risk of a t-shirt are rather low (basic product designs, materials, sizes) are decided internally.

Threadless uses prefabricated textiles, which makes the production process simple. Their business model focuses entirely on the co-design process and outsources printing to a local service provider. Nevertheless, the idea could be also successful by not using prefabricated products, but manufacturing the product from scratch once a winning design has been chosen. Indeed, the holding company behind Threadless is actually doing this. Motivated by its success in the fashion market, Threadless' founders have recently extended their categories to formal wear such as ties (NakedandAngry.com). Furthermore, Threadless has inspired a number of similar companies. While some have just cloned the idea (e.g. projectnvohk.com, look-zippy.com, buutvrij.com, lafraise.com – in total there have been more than 30 exact clones of the Threadless idea in the t-shirt market), companies like RYZwear, myfab.com or dreamheels.com have transferred the idea into another market segments (shoes, sneakers, furniture). The latter companies start the entire process just when the voting and commitment process has been finished.

10.2.2 Customer co-creation versus postponement and mass customization

Threadless' approach to customer co-creation in general and in particular its idea to determine the commitment of their customers before costly decisions are made on manufacture is just one of several recent strategies to increase the probability in meeting heterogeneous and fast-changing customer needs. Studies have shown that the forecasting accuracy can be improved dramatically after observing just 20% of the initial sales of an item.[10] Companies have reacted to this insight by delaying some activities, rather than starting them with incomplete information input, to cope better with the environmental uncertainty inherent in dynamic markets. In such a postponement strategy, manufacturing is split into two phases: an initial phase, where (generic) components are build-to-stock, and a second stage, where these components are transferred into the final product specification once more information about the market demand is available.[11] Connected with postponement, but different in nature, is mass customization. While in a postponement system the products are typically pre-defined by the supplier, with mass customization this process is reversed. It starts with customers co-designing their products, using a configuration system to specify their preferences. The individualized product is then manufactured on-demand.

Postponement and mass customization offer additional flexibility to minimize the new product development risk, but this flexibility does not come without costs. Both strategies require a redesign of the products and processes. This includes the creation of modular product family structures and often heavy investments in new flexible machinery equipment. For mass customization, also an elicitation system has to be in place to access the preferences of each individual customer and to transfer them into a precise product definition. On the operational level, postponement and mass customization imply costs of less efficient processing. As a result, mass customization and postponement are discussed broadly in the management literature, but few companies have implemented these strategies successfully today.[12]

Now compare Threadless' method to postponement and mass customization (see Fig. 10.1). Threadless has substituted conventional market research for deep continuous interactions with its customers. It does not ask its customers what they want to wear, but gives them a platform where they can express themselves and design these products. But most important, and contrary to earlier observations of customer or user driven innovation (see below), Threadless also transfers the decision process about what will be produced or not into the customer domain. Threadless provides its customer community with the capability to organize themselves and collect consensus over the most favorable upcoming products. Therefore, we call this method 'collective customer commitment'. It is important to remember that a new product design will only be finalized and go into production if enough customers pledge to purchase the design. In this way, market research expenditures are turned into early sales.

Threadless also needs less flexibility in its manufacturing system. Instead of investing in highly flexible manufacturing systems and dealing with individual custom designs, the company focuses its energy on motivating creative designers to submit new product ideas. Compared to mass customization, Threadless has

Postponement strategy	Mass customization	Collective customer commitment method
New product development by manufacturer (based on market research input)	Development of product architecture and customization options by manufacturer	Development of new product design by some (expert) customers
▼	▼	▼
Prefabrication of (some) components	Customer co-design process (elicitation)	Evaluation and refinement of design by manufacturer and customer community
▼	▼	▼
Access to better market information (based on market research input)	Placing of order by each individual customer	Presentation of selected design concepts and obtaining commitment of potential customers
▼	▼	▼
Final assembly of product variant	Custom (on-demand) manufacturing	Only if minimum lot size is pre-sold, (mass) production of product starts
▼	▼	▼
Mass distribution	Custom distribution	Mass distribution

10.1 The collective customer commitment method combines ideas of postponement and mass customization, but adds its own characteristics as well.

not to interact with individual customers and to run a complex on-demand manufacturing system. The costly configuration process in a mass customization system is substituted for an early involvement of some (expert) customers in the development and refinement of their ideas and pre-order by a larger group of customers. Likewise from the customers' perspective, the effort and risk of deciding about a custom design – mandatory in a mass customization configurator – is replaced by the security of peer-evaluated products.

10.2.3 When customer co-creation makes sense

Integrating customers in the innovation process and collecting customer purchase orders in advance of expenditures on detailed design and production may sound like the obscure idea of a small company in a niche market, but is becoming an increasingly popular approach with large companies as well. Indeed, in some markets this is the dominant way of generating business. Consider the real estate market. Here, condominiums are often sold like Threadless t-shirts; the developer will only start the construction when a given number of buyers have shown their willingness to purchase an apartment by placing a downpayment. But what has been a successful approach for very costly products, such as condos, in the past is passing downwards to fast-moving consumer commodities. We see two situations when the collective customer commitment method provides most value:

- to test really innovative products where little customer experience exists and thus market research is very fuzzy;
- to create perfectly matching products for rather small and very heterogeneous market segments.

Yamaha, a large manufacturer of musical instruments, employed the collective customer commitment method in the first situation. Yamaha's design team had envisioned an innovative electronic guitar, based on the feedback of frustrated, but lazy hobby musicians who wanted to play an instrument without practicing. The team came up with an instrument where, once fed with a song, small lights would tell the user where to press their fingers. This idea was breaking with the traditional design of a guitar and was considered too risky to be produced and developed in the conventional system. Thus, Yamaha used an existing user community to find out if there would be enough customer commitment for this design.[13] Users quickly drew on the idea and provided suggestions for improvements, such as adding an amplifier and making the device battery-powered. Once the final design was posted by Yamaha, the minimum order quantity was reached almost immediately, motivating Yamaha to produce this product. Currently it has sold more than 20 000 units, five times more than the average product in this category.

The second situation relates to a market where customer demand is very heterogeneous, a common situation today in many markets due to fast changing trends and more diverse needs.[14] Also the borders of formerly local markets are diminishing, and customer needs become geographically more broadly distributed. In heterogeneous and distributed markets, however, information about the demand for (new) products is distributed in an extremely diverse way, leading to large information asymmetries between individual customers and manufacturers. For manufacturers who want to provide an offering fitting exactly into such a market segment in order to exploit this differentiation opportunity promising high margins, it will become very costly to access all the required information.[15] If the knowledge of manufacturers about the needs of an emerging market is scarce and costly to achieve via conventional market research, user contributions will become a valuable source of innovation. The possibility of open contributions encourages self-screening by potential contributors. The capabilities of online interaction via the internet enable this process today for almost all product categories, independent of their overall market value. Thus, the strategy of powerful real estate developers to hedge their risk by pre-selling apartments can now be repeated for almost every product and by every manufacturer.

When discussing the specifics of customer co-creation, it is important to note that not everyone wants to actively participate in product development activities. Not all customers are highly motivated co-creators. Customers can decide about the degree of their involvement. At Threadless, most new designs are submitted by young professional designers, that is, users with typical lead user or trendsetting characteristics. They contribute not only because the monetary incentive of $2000 is higher than the average honorarium paid for a commissioned design by a conventional clothing company (about $300 to $500). Their main motivation is to get greater exposure on the professional design scene, a rather closed market that is difficult for newcomers to enter. The openness of Threadless' community makes it easy for designers to present their work and to get immediate feedback. But Threadless allows also pure hobbyists to submit a design as the screening activities by its community enable this openness at no risk and with no costs. Other users just comment on the submissions and propose amendments or additions. The majority of Threadless' users, however, just screens the proposals and contributes to the elicitation of demand by polling for the designs they like most. For these customers, browsing through the ideas is often a novel experience and a welcome change from traditional shopping activities.[16] They discover new potential products, exchange comments, and feel empowered by having the authority to make a favorite idea happen.

Beyond the motivation of participants, the contributors' knowledge also is essential to the entire process.[17] At Threadless, many of contributing designers are trained graphic designers. While everyone who submits a design has the same chance to win the contest, the expert knowledge of the designers gives them a clear advantage. From an economic point of view, bearing expert knowledge in the field of the co-creation process provides a cost advantage to the contributor. Another participant with less domain knowledge may be able to submit an equal contribution in the end, but at a much higher (development) cost. However, there are many examples at Threadless where also an outsider has submitted some highly creative and innovative designs. In these cases, often the intrinsic motivation of these hobbyists has counterbalanced their lack of experience. It may be exactly this openness to utilize expert knowledge on the one hand, but also to allow contributions from 'untrained' (and hence, unbiased) outsiders on the other, that explains the success of the Threadless model.

10.2.4 Implementing the collective customer commitment method

There are several benefits for manufacturers in implementing customer co-creation and the collective customer commitment method in particular. By creating an open line for their customers, manufacturers get access to ideas for new products or even complete designs. Supporting recent and potential customers in organizing themselves as a group and expressing commitment for a specific design turns market research expenditures into sales. Once this commitment is explicit, manufacturers can exploit this collective demand and serve the market very efficiently without the conventional costs of identifying this segment and the risk of developing and producing an unappealing offering.

However, we do not claim that customer co-creation and implementing the collective customer commitment method is always beneficial. Conventional product development and customer co-creation have to be seen as supplementary – not as substitutes. Successful innovation management is like any other management task, first of all, a decision about trade-offs, choosing what to do and what not to do. There will be contingency factors in favor of a manufacturer-dominated innovation process without any participation of the customer. But there is no doubt that customer integration matters in the new product development process of apparel companies. For the fashion and textile industry, a number of conditions have to hold so that customer co-creation makes sense in one particular category or for one particular company:

- Companies have to face uncertainty of demand. Especially in volatile markets influenced by fast moving fashion trends, integrating the customer in a different, much closer way makes sense. In the apparel market, obviously this makes no sense for classics or so called 'never out of stock' products. These are products where the uncertainty with regard to consumer preferences is rather low. In such markets, capabilities like manufacturing and managing a supply chain efficiently to guarantee availability and competitive prices are much more important than customer co-creation.
- The product has to be modular in such a way that it can be split in components that are pre-defined (and which in the best case could be pre-fabricated to reduce lead times) and others where customer co-creation can take place. This split reduces the complexity of the entire process and allows the external contributors to focus on just one aspect of the co-development. Splitting the product in such an 'internal' and 'external' way should start at the level of uncertainty about market demand. The components that are rather certain and bear a low planning risk (such as the basic t-shirt and size distribution, following the example of Threadless) will become pre-defined. Those where the companies face large demand uncertainty will be co-created with the consumers.
- Consumers have to be interested in the co-creation of this fashion product. This demands a specific level of involvement in the product. Next to high fashion, this may hold true, for example, for sports goods, but also 'functional' wear. Looking into the motivation of potential contributors is crucial for success.
- The company has to be able to create a community of contributors or to connect with an existing one. This often is the most challenging task. It took Threadless

more than five years to get its community running and working. Creating a community of contributors is, most of all, influenced by corporate culture. An important condition for success is the full disclosure of the entire process from initial consumer comments to final product commercialization. Often companies develop their products in secrecy, fearful of the prying eyes of competitors, for an ideal customer who may not actually exist. Co-creation builds on the integration of customers in an open innovation process. This demands an open, transparent development process, contrary to the conventional practice of keeping innovation closed and secret. Being able to make this mental shift is perhaps the most important condition for implementation.

Threadless has been able to make this cultural shift. But even further, it has tapped into a fundamental economic shift, a movement away from passive consumerism. Eventually, Threadless-like communities could form around industries as diverse as semiconductors, auto parts and toys. 'Threadless is one of the first firms to systematically mine a community for designs, but everything is moving in this direction,' Eric von Hippel has been quoted as saying in an interview.[18] This may or may not come to pass, but the lesson of Threadless is more basic. Its success demonstrates what happens when a company allows itself to be what its customers want it to be, when it makes something as basic and quaint as 'trust' a core competency. Threadless succeeds by asking more than any modern fashion company has ever asked of its customers – to design the products, to serve as the sales force, to become the employees. Its founders have pioneered a new kind of innovation. It does not require huge research budgets or creative brilliance – just a willingness to keep looking outward.

10.3   Notes

1 Balachandra and Friar (1997); Urban and Hauser (1993); Poolton and Barclay (1998); Redmond (1995); Tollin (2002).
2 Henkel and von Hippel (2005). Refer also to Adams *et al*. (1998); Bacon *et al*. (1994); Teas (1994).
3 The parent company of threadless.com is Chicago-based Skinnycorp.
4 Burke (1996) provides a good review of the inefficiencies of traditional market research.
5 Adams *et al*. (1998); Mahajan and Wind (1992).
6 Piller and Ihl (2010).
7 Franke and Piller (2003, 2004); Tseng *et al*. (2003); von Hippel (1998). A related and quite well researched method of customer co-creation is user idea contests. Several studies investigate those in a consumer goods setting, for example, Ebner *et al*. (2008); Piller and Walcher (2006); Sawhney *et al*. (2005). For a general literature review, refer to Rindfleisch and O'Hern (2008) and Piller and Ihl (2010).
8 A good review of research on customers as sources of innovation is provided by von Hippel (2005). Sawhney *et al*. (2003) show that these customers are often organized in communities by a manufacturer or intermediary. Piller *et al*. (2005) comment on the opportunities to perform co-design activities in a community.

9 Ogawa and Piller (2006). Elofson and Robinson (1998) describe a similar system called 'custom mass production'. Users first negotiate on a particular product design, find consensus about a solution that fits the desires of all, and then auction the resulting common idea to interested manufacturers.

10 Fisher and Raman (2001).

11 McCutcheon *et al.* (1994).

12 See with regard to postponement Gupta and Benjaafar (2004); Skipworth and Harrison (2004); with regard to customization Agrawal *et al.* (2001); Zipkin (2001); Salvador *et al.* (2009).

13 Yamaha teamed up with Engine, Inc., a competitor of Elephant Design (see note 9). Engine focuses on fashion items and the merchandizing of movie and comic characters (its 2004 sales topped ¥570 million). Registered users can submit 'please, make this' posts, i.e. ideas for new products, on its website tanomi.com (the name derives from the Japanese term *tanomikomu*, meaning requesting, referring both to the consumers' requests to produce a design and the manufacturers' request to purchase the product before production). Once copyright and production feasibility are cleared by a company board, the idea is published to the whole community for evaluation, together with a price and minimum order quantity for its commercialization. In addition, Engine offers other manufacturers the opportunity to post innovative product concepts directly to its community.

14 See Zuboff and Maxmin (2002) for an analysis of the reasons why markets are becoming more heterogeneous.

15 Von Hippel (2005: 72–75) calls these domains where large information asymmetries between individual users and manufacturers exist 'low-cost innovation niches', i.e. fields where information held locally by individual users strongly motivates them to contribute actively to a new development. With regard to this information transfer problem, see also von Hippel (1994) and Ogawa (1998).

16 On the internet, a growing number of websites serves this demand of innovation-seeking consumers (e.g., gizmodo.com, coolhunting.com or boingboing.net). They allow users, however, only to discover existing new products, but do not provide any open line to the manufacturers or product developers.

17 We gratefully acknowledge an anonymous reviewer for this comment.

18 Inc. Magazine, August 2008.

10.4 References

Adams, M.E., G.S. Day and D. Dougherty (1998). 'Enhancing new product development performance: an organizational learning perspective', *Journal of Product Innovation Management* 15 (September): 403–422.

Agrawal, M., T.V. Kumaresh and G. Mercer (2001). 'The false promise of mass customization', *McKinsey Quarterly* 38(3): 62–71.

Bacon, G., S. Beckman, D. Mowery, E. Wilson (1994). 'Managing product definition in high-technology industries', *California Management Review* 36 (Spring): 32–56.

Balachandra, R. and J.H. Friar (1997). 'Factors for success in R&D projects and new product introduction', *IEEE Transactions on Engineering Management* 44(3): 276–287.

Burke, R. (1996). 'Virtual shopping: breakthrough in marketing research', *Harvard Business Review* 74 (March-April): 120–129.

Ebner, W., J. Leimeister and H. Krcmar (2009). 'Community engineering for innovations: idea competitions as method to nurture virtual community for innovations', *R&D Management*, 39(4): 342–356.

Elofson, G. and W.N. Robinson (1998). 'Creating a custom mass-production channel on the internet', *Communications of the ACM* 41 (March): 56–62.

Fisher, M. and A. Raman (2001). 'Reducing the cost of demand uncertainty through accurate response to early sales', *Operations Research* 44 (January–February): 87–99.

Franke, N. and F. Piller (2003). 'Key research issues in user interaction with configuration toolkits in a mass customization system', *International Journal of Technology Management* 26(5): 578–599.

Franke, N. and F. Piller (2004). 'Toolkits for user innovation and design: exploring user interaction and value creation in the watch market', *Journal of Product Innovation Management* 21(6): 401–415.

Gupta, D. and S. Benjaafar (2004). 'Make-to-order, make-to-stock, or delay product differentiation? A common framework for modeling and analysis', IIE Transactions 36 (June): 529–546.

Henkel, J. and E. von Hippel (2005). 'Welfare implications of user innovation', *Journal of Technology Transfer* 30 (January): 73–88.

Mahajan, V. and J. Wind (1992). 'New product models: practices, shortcomings and desired improvements', *Journal of Product Innovation Management* 9 (June): 128–139.

McCutcheon, D.M., A. Raturi and J.R. Meredith (1994). 'The customization-responsiveness squeeze', *Sloan Management Review* 35 (Winter): 89–99.

Ogawa, S. (1998). 'Does sticky information affect the locus of innovation? Evidence from the Japanese convenience store industry', *Research Policy* 26 (July–August): 777–790.

Ogawa, S. and F. Piller (2006). 'Reducing the risks of new product development', MIT Sloan Management Review 47 (Winter): 65–72.

Piller, F. and C. Ihl (2010). *Open Innovation with Customers*. Raleigh, NC/New York: Lulu.com.

Piller, F., P. Schubert, M. Koch and K. Moeslein (2005). 'Overcoming mass confusion: collaborative customer co-design in online communities', Journal of Computer-Mediated Communication 10(4) [online journal: jcmc.indiana.edu/vol10/issue4/piller.html].

Piller, F. and D. Walcher (2006). 'Toolkits for idea competitions: a novel method to integrate users in new product development', *R&D Management* 36(3): 307–318.

Poolton, J. and I. Barclay (1998). 'New product development from past research to future applications', *Industrial Marketing Management* 27(3): 197–212.

Redmond, W.H. (1995). 'An ecological perspective on new product failure: the effects of competitive overcrowding', *Journal of Product Innovation Management* 12 (June): 200–213.

Rindfleisch, A. and M. O'Hern (2008). 'Customer co-creation: a typology and research agenda' (University of Wisconsin Working Paper).

Salvador, F., M. de Holan and F. Piller (2009). 'Cracking the code of mass customization', *MIT Sloan Management Review* 50(3): 70–79.

Sawhney, M., E. Prandelli and G. Verona (2003). 'The power of innomediation', *Sloan Management Review* 44 (Winter): 77–82.

Sawhney, M., G. Verona and E. Prandelli (2005). 'Collaborating to create: the internet as a platform for customer engagement in product innovation', *Journal of Interactive Marketing* 19(4): 4–17.

Skipworth, H. and A. Harrison (2004). 'Implications of form postponement to manufacturing: a case study', *International Journal of Production Research* 42(10): 2063–2081.

Teas, R.K. (1994). 'Expectations as a comparison standard in measuring service quality: an assessment of a reassessment', *Journal of Marketing* 58 (January): 132–139.

Tollin, K. (2002). 'Customization as a business strategy: a barrier to customer integration in product development', *Total Quality Management* 13 (July): 427–439.

Tseng, M., T. Kjellberg and S. Lu (2003). 'Design in the new e-commerce era', *Annals of the CIRP* 52(2): 509–519.

Urban, G. and J. Hauser (1993) Design and Marketing of New Products (2nd edn). Englewood Cliffs, NJ: Prentice Hall.

von Hippel, E. (1994). 'Sticky information and the locus of problem solving', *Management Science* 40 (April): 429–439.

von Hippel, E. (2005). Democratizing Innovation. Cambridge, MA: The MIT Press.

Zipkin, P. (2001). 'The limits of mass customization', *Sloan Management Review* 42 (Spring): 81–87.

Zuboff, S. and J. Maxmin (2002). *The Support Economy: Why corporations are failing individuals and the next episode of capitalism*. London: Viking Penguin.

11

The development and marketing of SilverClear®

L. HORNE, University of Manitoba, Canada
and B. ROSE, TransTex Technologies Inc., Canada

Abstract: This chapter documents TransTex Technologies, a small company's experience in developing and marketing an antibacterial finish called SilverClear®. It describes the market segments served by this company and the costs involved in exporting its products to the US. This case is an example of the importance of recognizing the opportunities and barriers presented by the regulatory environment.

Key words: antibacterial finish, marketing, regulatory environment.

11.1 Introduction

In 2008 the Office of the Auditor General of the Province of Ontario, Canada, released a report called 'Prevention and control of hospital-acquired infection'. This report cited a 2003 Canadian study that estimated that nosocomial infections could cause 8000 deaths annually (Office of the Auditor General of Ontario, 2008). Nosocomial infection is a serious health concern for many countries in the world. As governments search for ways to harness infection control in public places, the textiles sectors of many countries have successfully utilized textiles as one of many solutions to the problem.

Textiles with biocidal attributes are readily available in the marketplace nowadays. Silver is one of the agents used in producing biocidal textiles. The purpose of this chapter is to describe the processes of a company experienced in developing and marketing an antimicrobial finish. Specifically, this case study is intended to show the role of regulatory agencies in the development of a small business, TransTex Technologies Inc., that serves the medical textiles market. The setting is Quebec, a province situated in Eastern Canada. Quebec is also a major textile manufacturing centre in Canada. In 2010 exports of textiles and non-apparel textile products from this province amounted to over US$700 million (Industry Canada, 2010).

11.1.1 What is SilverClear®?

SilverClear® is a silver-based, colourless coating manufactured by TransTex Technologies Inc., located in Quebec, Canada. It is made of 'slightly soluble, silver salt crystals that can be applied in different finishing/coating processes such

190

as spraying, dip-coating, padding and thin film deposition to form a durable antimicrobial/biocidal polymer coating' (Tessier, 2008: 17). This coating has antibacterial (inhibiting growth of bacteria) and bactericidal (destroying bacteria) properties, and has been proven to be effective in neutralizing or destroying bacteria such as S. aureus, P. aeruginosa, E. coli and C. difficile. (SilverClear™, Proven Efficiency, n.d.). A distinct attribute of this coating is that it is colorless, hence it can be applied to fibers, yarns, papers, woven or non-woven materials, composites or plastics. (SilverClear™, Versatile, n.d.).

11.1.2 End uses for SilverClear®

Since its commercialization, SilverClear® has been applied to textiles for various markets including healthcare, medical, industrial and hospitality, protection, and clothing. (SilverClear™, Markets, n.d.). Demand is highest in three end uses – critical wound healing, post-radiation care and oral surgery.

- *Wound healing* There is a range of prescription dressings for institutional uses. These dressings are sold to hospitals in bulk for use in the treatment of severe burns and open wounds that do not respond well to other treatments.
- *Radiation treatments* There is a range of garments for the upper and lower body made from stretch fabrics for post-radiation care.
- *Oral surgery* This range comprises pallet grafts for oral surgery. This line of products has received favourable response from a number of dentists. Research is being undertaken to validate its efficacy.

In addition to critical wound healing, post-radiation care and oral surgery, SilverClear® is also used in producing dressings or post-operative garments that accelerate the healing of scars. Finally, TransTex Technologies also provides custom fitted garments made from fabrics treated with SilverClear® for humans as well as animals through a network of small manufacturers.

11.1.3 Development of SilverClear®

SilverClear® was developed in the 1990s in response to other silver-treated antimicrobial textiles that were very costly due to high silver content (some of the materials could contain as much as 25% silver). Bernard Rose believed that there had to be a less costly method of imparting silver in textiles. In the late 1990s Mr Rose successfully enlisted a research and development organization to collaborate with him to develop a cost-efficient antibacterial material using silver.

The first research and development effort, which lasted about three years, was to explore the use of plasma technology to impart silver. This process resulted in limited success because only one side of a fabric could be coated at a time, and the flexibility of the material left gaps in the treated area. The second research and development effort, which lasted about one year, was a chemical approach. This chemical approach yielded a milky-white liquid solution that contained less than one per cent silver. When textiles treated with this solution were tested using the American Association of Textile Chemists and Colorists (AATCC) methods, the antibacterial performance was very effective. Due to the extreme surface-to-mass ratio of the particles involved, a very small amount of silver is needed to impart antibacterial properties, which makes the application of this treatment much less costly. For example, a glove treated with SilverClear® could be sold to hospitals for CAD$92, while an earlier generation glove treated with agents with much higher silver content could cost hospitals as much as CAD$900.

In the beginning, in addition to wound care end uses, TransTex Technologies marketed SilverClear® for the treatment of undergarments, socks, and shoe linings for odor control and the prevention of other problems related to microbes and bacteria. Business opportunities were then pursued in the export market and other applications, such as bedding, latex paint, cat litter and industrial applications where the anti-fungal properties of SilverClear® could be beneficial.

Mr Rose aimed to have focussed on expanding the use of SilverClear® in the medical market by the middle of the first decade of the twenty-first century. However, as TransTex Technologies is a small company, direct competition with large multinationals is not a practical strategy. Therefore, Mr Rose decided to sell to the Canadian market. In order to sell medical textiles in Canada, products treated with SilverClear® needed to be registered as a medical device in Canada.

11.2   The medical device industry in Canada

The medical device industry in Canada 'consists of firms that produce a wide range of products used for diagnosis and treatment of ailments, and which include the following: medical, surgical and dental equipment (including electromedical equipment and related software), furniture, supplies and consumables, orthopaedic appliances, prosthetics and diagnostic kits, reagents, and equipment. Firms that are active only in distribution are not included in the profile' (Industry Canada, n.d.). Medical devices made and marketed in Canada must comply with the Canadian Food and Drugs Act.

According to Statistics Canada, laboratories, other companies and universities were sources of spin-offs for ten per cent of medical device firms in Canada (Statistics Canada, 2002). From 2000 to 2007 the compound annual growth rate was 5.8% and this market was valued at $7.1 billion in 2007 (Industry Canada, n.d.).

11.2.1 The regulatory arena

In Canada, depending on the product and its end use, the registration of an antibacterial treatment for textile products could fall under the authority of either the Medical Devices Directorate (MDD) or the Pest Management Regulatory

Agency (PMRA). Some products and end uses, such as wound care dressings, which apply directly to the protection of the person, fall naturally within the MDD. However, antibacterial treatment for textiles for end uses such as upholstery falls under the jurisdiction of the PMRA because the antibacterial protection is directed to a product, not a person. In other words, when the treatment is for the protection of the person and/or relief of symptoms related to the person, the product falls under the authority of the MDD. If the treatment is for the protection of a product, it falls under the authority of the PMRA.

11.2.2 The Medical Devices Directorate (MDD)

The Medical Devices Directorate registers and authorizes the use of medical devices as outlined in the Canadian Food and Drugs Act and its regulations. The cost of registration is comparable to the cost of registration with the Food and Drug Administration (FDA) in the United States, and with similar agencies in other industrialized nations.

Medical devices in Canada are grouped into four categories based on the level of risk associated with their use (Industry Canada, 2005). Class 1 medical devices present the lowest potential risk; products such as thermometers and medical stockings fall under this classification. Class 4 medical devices, such as pacemakers and heart valves, present the highest potential risk. Manufacturers of Class 1 devices need not list their devices with Health Canada, but must hold a 'Medical Device Establishment License' and comply with the safety standards set forth in the Medical Device Regulations. In contrast, Classes 2, 3 and 4 devices receive increasingly vigorous reviews, and must be listed with Health Canada's Medical Devices Active Licensing Listing (MDALL) to be sold in Canada.

Wound dressings treated with SilverClear® fall into the description of a Class 2 medical device. To register SilverClear® wound dressings, Mr Rose enlisted the expertise of a medical device registration consultant to help him navigate the bureaucracy involved in this process. Companies that choose to use a consultant can expect to spend $25–30 000 to register a Class 2 medical device. The time required to obtain a Class 2 licence is typically two to three months after submission of registration.

In addition to meeting the Medical Device Regulations, the Canadian government also requires that medical devices be manufactured in accordance with ISO 13485 standards. A manufacturer must show regulators substantive scientific evidence of a product's safety, efficacy and quality as required by the regulations stipulated in the Food and Drugs Act.

In January of 2005 Transtex Technologies received its licence to manufacture and distribute over 200 SilverClear® items, including dressings and garments for wounds and burns. Once the products were registered, they could be prescribed by medical doctors in Canada. To distribute SilverClear® products, Mr Rose set up a distribution channel and worked with the health care ministries in the various provinces to list products in the appropriate medical device registry.

11.2.3 The Pest Management Regulatory Agency

As the range of products treated with SilverClear® continued to expand, the application was no longer limited to products that would be used directly to treat patients. Two developments in enlarging the market for SilverClear® products presented substantial challenges to the company. The first challenge presented itself when TransTex Technologies applied for a licence to manufacture and distribute curtains treated with SilverClear® for use in hospitals. Mr Rose found himself facing a registration process that was very different from registering a Class 2 medical device.

TransTex Technologies wanted to produce curtains treated with SilverClear® because textile products such as uniforms, curtains and upholstery are examples of numerous sources of cross contamination in hospitals and health care facilities. These textile products could harbor bacteria, viruses and other pathogens that could infect patients whose health has already been compromised. A study of contamination of hospital curtains in a hospital in Cleveland, US showed that 20% of the curtains tested positive for VRE, 22% tested positive for MRSA, and 4% tested positive for C. difficile (Trillis *et al.*, 2008). As doctors and nurses habitually pull the curtains that separate beds in hospital rooms, they would brush against the curtains and risk contaminating either themselves and/or the curtains. In Mr Rose's mind, treating the curtains with an antimicrobial finish would result in the protection of humans.

As Mr Rose delved deeper into the registration process, he discovered that, according to the guidelines of the Food and Drugs Act, wound dressings treated with SilverClear® would be interpreted as medical devices because the products are directly applied to patients, rendering protection for humans. However, curtains treated with SilverClear® would be interpreted as non-medical use of antimicrobial treatments because the treated curtains were not used directly on humans to protect them from bacterial infection. Within that interpretation, the SilverClear® treatment would be considered pesticidal; its registration would fall under the jurisdiction of the PMRA.

Mr Rose discovered that the cost of registering curtains treated with SilverClear® with PMRA was far higher than registering a Class 2 medical device with MDD. The cost to register SilverClear® treated curtains and other medical textiles with PMRA could be as much as CAD$265 000, plus the substantial cost of tests and consulting fees. This could be equivalent to as much as 10 or 12 times the potential profit to be made in the relatively small Canadian market in the first year. Furthermore, the time required to obtain such a licence could be as long as 32 months including submitting a registration application. The product development cost and the long time involved in registration were impediments to a small company such as TransTex Technologies, operating in the relatively small Canadian market.

To expand its market reach, TransTex Technologies began to investigate the feasibility of exporting SilverClear® products to the United States, where the treatment of textile materials for non-medical applications fell under the authority of the Environmental Protection Agency (EPA). The basic cost of registering SilverClear® products with EPA was US$27 000 (compared with CAD$ 265 000 with PMRA in Canada); another US$125 000 would be spent on testing and consulting fees. Although US$27 000 was a substantial amount of money, a provision in the EPA registration allowed small and medium sized companies to enjoy a cost relief that effectively lowered the registration cost for TransTex Technologies by about 90%. The length of time from submission of a registration to receiving a said registration was dependent on the complexity of the product and whether EPA reviewers were familiar with similar products.

11.2.4 Lessons learned

Mr Rose's experience with the regulatory agencies revealed two barriers for small Canadian companies such as TransTex Technologies to overcome when they want to serve the Canadian market. The high cost of registering products under non-medical applications resulted in a very high product development cost, which would inevitably be reflected in the product's selling price. This will preclude small companies that are capable of creating innovative products that meet the highest standards from competing with imported textiles treated with antimicrobials that have no proof of efficacy.

The second issue pertained to the interpretation of medical versus non-medical applications of antibacterial treatments. Mr Rose's argument was that the purpose of treating curtains with an antibacterial agent was not to protect the curtain but to protect the people, who, for one reason or another, found themselves in a hospital environment. Hence, the registration of such a product should not be processed through the lens of pest management.

11.3   The importance of market access

In a report released in April 2011 by the Medical Devices Innovation Institute in Canada, the medical devices sector is projected to grow in response to the health-related implications of an aging population. Although there were no specific data on the share of medical textiles to the entire medical devices sector, some types of textiles, such as those treated with SilverClear®, are classified as medical devices. In this report, 'the lack of harmonization with other (internal or external) jurisdiction' is identified as one of several barriers to the growth of the medical textiles sector (Medical Devices Innovation Institute, 2011: 6). This case study is an example of jurisdictions as barriers to growth for TransTex Technologies.

Fortunately, Mr Rose has been actively engaged in dialogues with various agencies to arrive at solutions that ultimately would enable Canadian companies to supply safe and effective medical textile products to their domestic market. For TransTex Technologies, its access to the US market received a promising boost in March 2010 when SilverClear® was successfully registered with that country's Environmental Protection Agency.

11.4   References

Industry Canada (n.d.). Canadian medical device industry. Available at http://www.ic.gc. ca/eic/site/md-am.nsf/vwapj/EnglishMedicalDeviceIndustryprofile_Oct26_2009. pdf/$FILE/EnglishMedicalDeviceIndustryprofile_Oct26_2009.pdf

Industry Canada (2005). Quality system requirements for medical devices reference guide for manufacturers selling medical devices in Europe, Canada and the United States. Available at http://dsp-psd.pwgsc.gc.ca/collection_2007/ic/Iu44-23-2005E.pdf

Industry Canada (2010). Trade data online. Available at http://www.ic.gc.ca/sc_mrkti/tdst/ tdo/tdo.php#tag

Medical Devices Innovation Institute (2011, April). Medical devices challenges and opportunities for enhancing the health and wealth of Canadians. Available at http:// www.medec.org/en/content/report-medical-devices-challenges-and-opportunities

Office of the Auditor General of Ontario (2008, September). Prevention and control of hospital-acquired infections. Available at http://www.auditor.on.ca/en/reports_en/hai_ en.pdf

Statistics Canada (2002). Survey of the medical devices industry. Available at http://www. statcan.gc.ca/cgi-bin/imdb/p2SV.pl?Function=getSurvey&SDDS=2947&lang=en&db= imdb&adm=8&dis=2

SilverClear™ (n.d.). Markets. Available at http://www.silverclear.ca/index_main.html

SilverClear™ (n.d.). Proven efficiency. Available at http://www.silverclear.ca/html/data.html

SilverClear™ (n.d.). Versatile. Available at http://www.silverclear.ca/html/versatile.html

Tessier D. (2008, January). SilverClear™: An outstanding silver technology. *The Textile Journal*, 125: 14–18.

Trillis III, F., Eckstein, E. C., Budavich, R., Pultz, M. J. and Donskey, C. J. (2008). Contamination of hospital curtains with healthcare-associated pathogens. *Infection Control and Hospital Epidemiology*, 29(11): 1074–1076.

Index